my **revision** notes

AQA **A2**

BIOLOGY

Mike Boyle

Hodder Education, an Hachette UK company, 338 Euston Road, London NW1 3BH

Orders
Bookpoint Ltd, 130 Milton Park, Abingdon, Oxfordshire OX14 4SB
tel: 01235 827827
fax: 01235 400401
e-mail: education@bookpoint.co.uk
Lines are open 9.00 a.m.–5.00 p.m., Monday to Saturday, with a 24-hour message answering service. You can also order through the Hodder Education website: www.hoddereducation.co.uk

© Mike Boyle 2013
ISBN 978-1-4441-7976-7

First printed 2013
Impression number 5 4 3 2 1
Year 2017 2016 2015 2014 2013

All rights reserved; no part of this publication may be reproduced, stored in a retrieval system, or transmitted, in any form or by any means, electronic, mechanical, photocopying, recording or otherwise without either the prior written permission of Hodder Education or a licence permitting restricted copying in the United Kingdom issued by the Copyright Licensing Agency Ltd, Saffron House, 6–10 Kirby Street, London EC1N 8TS.

Cover photo reproduced by permission of Kletr/Fotolia

Typeset by Datapage (India) Pvt. Ltd.
Printed in India

Hachette UK's policy is to use papers that are natural, renewable and recyclable products and made from wood grown in sustainable forests. The logging and manufacturing processes are expected to conform to the environmental regulations of the country of origin.

P2186

Get the most from this book

Everyone has to decide his or her own revision strategy, but it is essential to review your work, learn it and test your understanding. These Revision Notes will help you to do that in a planned way, topic by topic. Use this book as the cornerstone of your revision and don't hesitate to write in it — personalise your notes and check your progress by ticking off each section as you revise.

☑ Tick to track your progress

Use the revision planner on pages 4 and 5 to plan your revision, topic by topic. Tick each box when you have:

- revised and understood a topic
- tested yourself
- practised the exam questions and gone online to check your answers and complete the quick quizzes

You can also keep track of your revision by ticking off each topic heading in the book. You may find it helpful to add your own notes as you work through each topic.

My revision planner

Unit 4 Populations and environment

		Revised	Tested	Exam ready
1 The dynamic equilibrium of populations				
8	Populations and ecosystems	☐	☐	☐
9	Investigating populations	☐	☐	☐
11	Variation in population size	☐	☐	☐
12	Human populations	☐	☐	☐
14	An introduction to statistics	☐	☐	☐

Populations and ecosystems

Biotic and abiotic factors Revised

A **population** is all the individuals of a particular species that can interbreed. A **community** is all the populations in the ecosystem, including plants, animals and **decomposers**. This section is about the ways in which organisms are affected by environmental factors, which can be either biotic or abiotic.

The most common **biotic** (living) factors include:

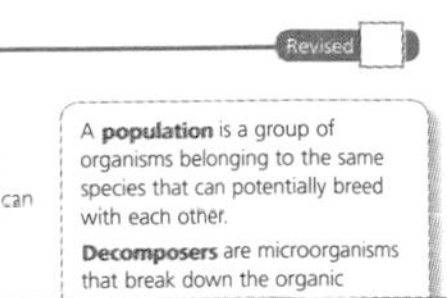

A **population** is a group of organisms belonging to the same species that can potentially breed with each other.

Decomposers are microorganisms that break down the organic

Features to help you succeed

Examiner's tips and summaries

Expert tips are given throughout the book to help you polish your exam technique in order to maximise your chances in the exam.

The summaries provide a quick-check bullet list for each topic.

Typical mistakes

The author identifies the typical mistakes candidates make and explains how you can avoid them.

Revision activities

These activities will help you to understand each topic in an interactive way.

Now test yourself

These short, knowledge-based questions provide the first step in testing your learning. Answers are at the back of the book.

Exam practice

Practice exam questions are provided for each topic. Use them to consolidate your revision and practise your exam skills.

Definitions and key words

Clear, concise definitions of essential key terms are provided on the page where they appear. Key words you need to know are highlighted in bold throughout the book.

Online

Go online to check your answers to the exam questions and try out the extra quick quizzes at **www.therevisionbutton.co.uk/myrevisionnotes**

My revision planner

Unit 4 Populations and environment

Unit 5 Control in cells and organisms

Exam practice answers and quick quizzes at www.therevisionbutton.co.uk/myrevisionnotes

Countdown to my exams

6–8 weeks to go

- Start by looking at the specification — make sure you know exactly what material you need to revise and the style of the examination. Use the revision planner on pages 4 and 5 to familiarise yourself with the topics.
- Organise your notes, making sure you have covered everything on the specification. The revision planner will help you to group your notes into topics.
- Work out a realistic revision plan that will allow you time for relaxation. Set aside days and times for all the subjects that you need to study, and stick to your timetable.
- Set yourself sensible targets. Break your revision down into focused sessions of around 40 minutes, divided by breaks. These Revision Notes organise the basic facts into short, memorable sections to make revising easier.

Revised ☐

4–6 weeks to go

- Read through the relevant sections of this book and refer to the examiner's tips, examiner's summaries, typical mistakes and key terms. Tick off the topics as you feel confident about them. Highlight those topics you find difficult and look at them again in detail.
- Test your understanding of each topic by working through the 'Now test yourself' questions in the book. Look up the answers at the back of the book.
- Make a note of any problem areas as you revise, and ask your teacher to go over these in class.
- Look at past papers. They are one of the best ways to revise and practise your exam skills. Write or prepare planned answers to the exam practice questions provided in this book. Check your answers online and try out the extra quick quizzes at **www.therevisionbutton.co.uk/myrevisionnotes**
- Try different revision methods. For example, you can make notes using mind maps, spider diagrams or flash cards.
- Track your progress using the revision planner and give yourself a reward when you have achieved your target.

Revised ☐

One week to go

- Try to fit in at least one more timed practice of an entire past paper and seek feedback from your teacher, comparing your work closely with the mark scheme.
- Check the revision planner to make sure you haven't missed out any topics. Brush up on any areas of difficulty by talking them over with a friend or getting help from your teacher.
- Attend any revision classes put on by your teacher. Remember, he or she is an expert at preparing people for examinations.

Revised ☐

The day before the examination

- Flick through these Revision Notes for useful reminders, for example the examiner's tips, examiner's summaries, typical mistakes and key terms.
- Check the time and place of your examination.
- Make sure you have everything you need — extra pens and pencils, tissues, a watch, bottled water, sweets.
- Allow some time to relax and have an early night to ensure you are fresh and alert for the examination.

Revised ☐

My exams

A2 Biology Unit 4

Date: ....

Time: ....

Location: ....

A2 Biology Unit 5

Date: ....

Time: ....

Location: ....

The assessment objectives

For both AS and A2 biology, you will be tested on three assessment objectives (AOs):

- *AO1: Knowledge and understanding of science and of How Science Works* — these are straightforward factual questions, read them as 'Give us the facts, tell us what you know.' There are not as many of these questions as you may like; instead, there are a lot more of AO2.
- *AO2: Application of knowledge and understanding of science and of How Science Works* — these questions typically give you unfamiliar examples, data and graphs so you have to apply the knowledge you have. This is what good scientists can do. You can do it too, but it takes a bit of practice.
- *AO3: How Science Works* — these questions test practical skills or the skills you will need when designing and evaluating investigations. They also involve dealing with data, including statistics.

Differences between AS and A2

There are three significant differences between AS and A2:

- There are more AO2 questions (see the table below). The first A2 exams in 2010 made the national news because students thought they were far too hard and voiced their opinions via social media. In response, the exams were made a little easier, but the balance of assessment objectives stays the same. Most of the questions contain unfamiliar data and you will be expected to apply the principles you have learnt. The examiners deliberately choose unfamiliar examples so it's the same for everyone.
- You are expected to know about statistics. In the Unit 6 exam you will be expected to generate some data, choose a statistical test (from a choice of three), do the calculations and interpret the results. You won't get any long calculations to do and you can usually use a calculator or computer to help. With stats, the bottom line is always the same: what is the *probability* that these results are significant or just down to chance?
- There is a longer exam at the end (in Unit 5), which includes an essay. You have a choice of one from two, and you will need to write for about 35 minutes, which for most people is about three sides of A4. The essays are set on ideas and concepts that require you to gather knowledge from more than one unit and you cannot get full marks unless you include a couple of points that are beyond the specification, to show evidence of wider study.

The table shows how the three assessment objectives contribute to your final score. Note that there are more AO2 questions at A2.

Breakdown of marks

Assessment objective	Unit weighting (%)						Overall weighting of AOs (%)
	Unit 1	Unit 2	Unit 3	Unit 4	Unit 5	Unit 6	
AO1	7	9	2	5	6	1	30
AO2	7	9	1	8	13	2	40
AO3	3	5	7	4	4	7	30
Overall weighting of units	17	23	10	17	23	10	100

1 The dynamic equilibrium of populations

Populations and ecosystems

Biotic and abiotic factors

Revised

A **population** is all the individuals of a particular species that can interbreed. A **community** is all the populations in the ecosystem, including plants, animals and **decomposers**. This section is about the ways in which organisms are affected by environmental factors, which can be either biotic or abiotic.

The most common **biotic** (living) factors include:

- food supply
- **predation**
- competition for mates and nesting sites
- pathogens such as bacteria, viruses, fungi and parasites

The **abiotic** (non-living) environment includes:

- temperature
- light
- inorganic nutrients such as nitrate and phosphate
- carbon dioxide and oxygen levels
- pH
- humidity

There are also many other factors that affect organisms depending on their **habitat**.

In the AS course we looked at **diversity**. As a general rule:

- **Ecosystems** that have a **high diversity** have a favourable abiotic environment. For example, rainforests have plenty of sunlight, a favourable temperature, lots of water and a steady climate without unfavourable seasons. In these conditions, biotic factors tend to dominate an organism's life. There are thousands of different **niches** for different species.
- In contrast, in ecosystems with a **low diversity** such as deserts and arctic regions, abiotic conditions dominate. Not many organisms can survive low temperature and/or a lack of available water. Sometimes there is no water available at all and sometimes it is not available because it is frozen or too salty.

A **population** is a group of organisms belonging to the same species that can potentially breed with each other.

Decomposers are microorganisms that break down the organic compounds in dead plants and animals and release carbon dioxide and simple inorganic ions.

Predation occurs when one organism eats another.

An **ecosystem** is a natural unit consisting of living and non-living components. Examples include a coral reef, freshwater lake or temperate forest.

A **niche** is a concept that describes an organism's role in the ecosystem. Each species occupies a particular niche because it is adapted to a particular set of biotic and abiotic factors. No two species occupy exactly the same niche. Think of niche as an 'opportunity'.

Now test yourself

Tested

1 Distinguish between the terms *population* and *community*.
2 Oak trees are a common woodland species. List three biotic and three abiotic factors that might affect an individual oak tree.

Answers on p. 110

Investigating populations

Practical techniques

Revised

Investigating ecosystems involves **random and systemic sampling** in order to obtain **quantitative data**. You cannot possibly measure all that there is to measure, so the aim is to make sure that your sample is representative of the data as a whole.

Quadrats

Quadrats are frames that are used for comparing two areas of land, such as north- and south-facing walls or mown and un-mown fields. The species **abundance** and **distribution** can be measured, but this method is usually limited to plants, lichens and slow-moving animals such as snails or limpets.

How to use quadrats:

1 Map out the areas you want to study. This can be done virtually or actually (with pins and string, for example, or with two tape measures set at right angles).
2 Number each square or work out a coordinate system (like a chess board, perhaps A5, B6 etc. — see Figure 1.1).
3 Select numbers to sample. This must be done at random to **avoid bias**. Acceptable ways to do this include using a random-number generator, a telephone directory or a book of random numbers (yes, they do exist).
4 Take readings at the selected quadrats. What you actually do depends on the organisms in the sample. You could count the individuals to work out their **frequency** or estimate their **percentage cover**, or you could rate a particular species as abundant, common, frequent, occasional or rare (the ACFOR scale).

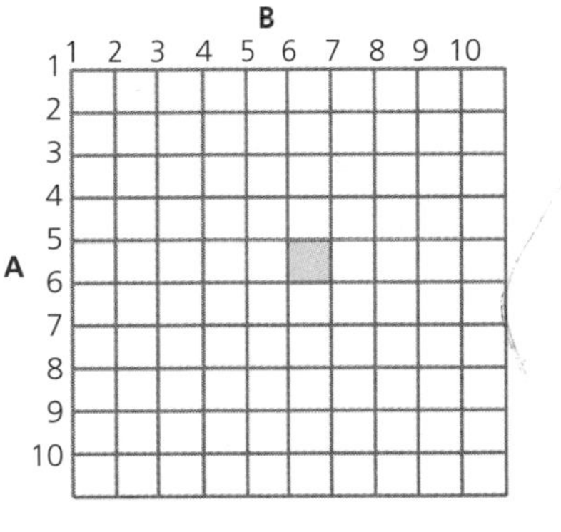

Figure 1.1 A grid with a quadrat positioned at A5, B6

Choosing a sample at random **avoids bias**, which means *without conscious choice*.

Transects

Transects are lines that are used to show change from one area to another. They can be used on their own or with quadrats. Situations where you might use transects include rocky shores, sand dunes and rivers. Different types of transect include:

- point transects — simply record the species present at points along the line
- belt transects — place a quadrat continuously alone the line and record the organisms in it
- interrupted belt transect — for longer surveys, a quadrat can be placed at regular intervals, say every 5 or 10 m (see Figure 1.2)

Examiner's tip

Remember that transects use systematic sampling, whereas quadrats use random sampling.

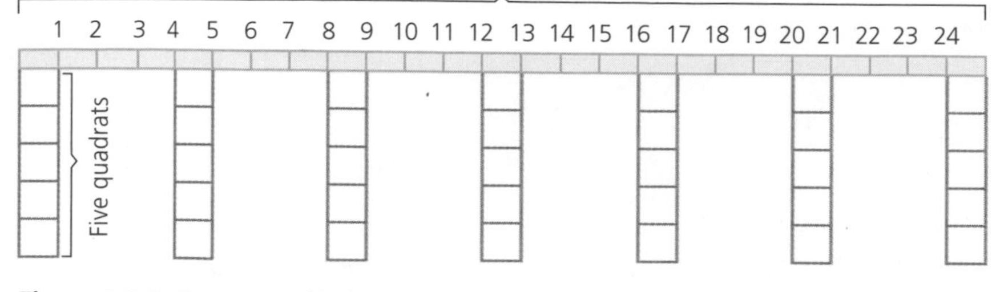

Figure 1.2 An interrupted belt transect. One or more quadrats can be used at every interval

Now test yourself Tested

3 Which type of sampling technique would be most suitable for investigating each of the following?
  (a) the growth of moss and lichens on north- and south-facing tree trunks
  (b) the distribution of organisms down a rock shoreline
  (c) the distribution of organisms along a sand dune from the sea

Answers on p. 110

Mark-release-recapture

Animal populations can be measured using **mark-release-recapture**. To estimate the population of animal species, the process is as follows:

1. Catch the animals. Pitfall traps work for small species like beetles, whereas Longworth traps are good for small mammals.
2. Mark them in a way that does not harm them, restrict their mobility or make them more visible to predators. A spot of paint works for snails, whereas a small patch of fur can be clipped when marking mammals.
3. Let them go and give them enough time to mix with the rest of the population.
4. Set traps again — keeping all methods and timings the same — and note how many are marked.

Examiner's tip

There are many different ways of trapping animals and you are not expected to know all of them. You can suggest improvisations in any given situation, such as pooters for sucking up small invertebrates, nets for butterflies and grasshoppers and kick samples for river or stream organisms. You can even beat tree branches with sticks and collect the insects that fall in a tray, sheet or an old umbrella.

The population can be estimated using the formula:

$$\text{population} = \frac{\text{number in first sample} \times \text{number in second sample}}{\text{number marked in the second sample}}$$

For example, in an investigation into a population of wood mice the following data were obtained:

First sample = 75 mice captured, marked and released

Second sample = 81 mice captured, of which 8 were marked

Therefore, the population is calculated as:

$$\frac{75 \times 81}{8} = 759 \text{ (rounded down)}$$

Now test yourself Tested

4 To estimate the population of periwinkles in a harbour, 200 individuals were collected and marked with a spot of paint. The next day another 200 were collected, of which 23 were already marked. Estimate the population.

Answer on p. 110

Variation in population size

Size of population

Revised

All populations increase in size when conditions are favourable and stabilise or decrease when things are unfavourable. The size of any population is limited by a complex **interaction** of factors, including:

- **interspecific competition**, which is between individuals of different species
- **intraspecific competition**, which is between individuals of the same species
- **predation**
- **abiotic factors**

Population growth curves

Revised

Starting with a small number of organisms, perhaps even one, population growth usually follows a particular pattern, as shown in Figure 1.3.

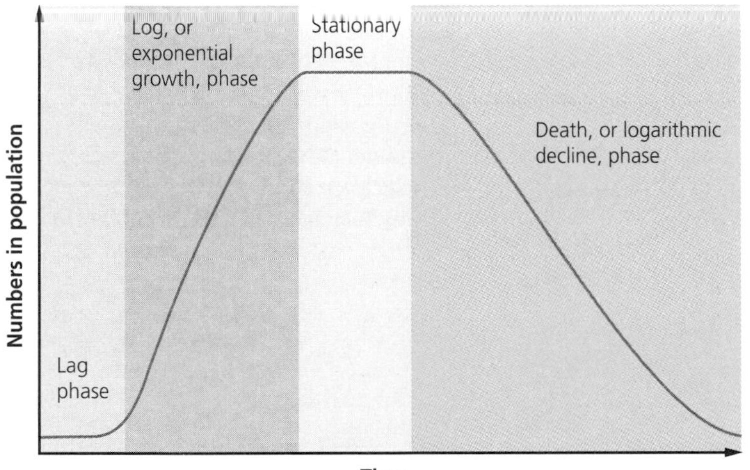

Figure 1.3 A population growth curve showing the main phases

Phase 1 Lag phase. Growth is slow for a variety of reasons that depend on the organism concerned. Bacteria need time to synthesise and secrete the right enzymes, whereas more complex organisms need to grow to sexual maturity.

Phase 2 Log phase. Growth is rapid because there are no **limiting factors**. Growth is exponential (2, 4, 8, 16, 32 etc.).

Phase 3 Stationary phase. Growth slows and stops. No matter what the situation, no population can go on expanding indefinitely. There are always limiting factors including lack of food, accumulation of waste and more predators.

Phase 4 Death phase. This phase depends on the situation. In a closed system, such as yeast fermenting in a bottle, the food will run out, waste will accumulate and all the individuals will die. In an open system — one in which there is an input of food or energy — the population settles at a sustainable level, known as the **carrying capacity**. This is the population that can be supported in any given situation.

Predators and prey

Revised

A certain population of prey animals can support a certain population of predators. A clear example of this is the snowshoe hare and the lynx (Figure 1.4). There are two reasons why this is a good example: there are no other species involved in this food chain and there are clear records of population changes from fur trappers in Canada.

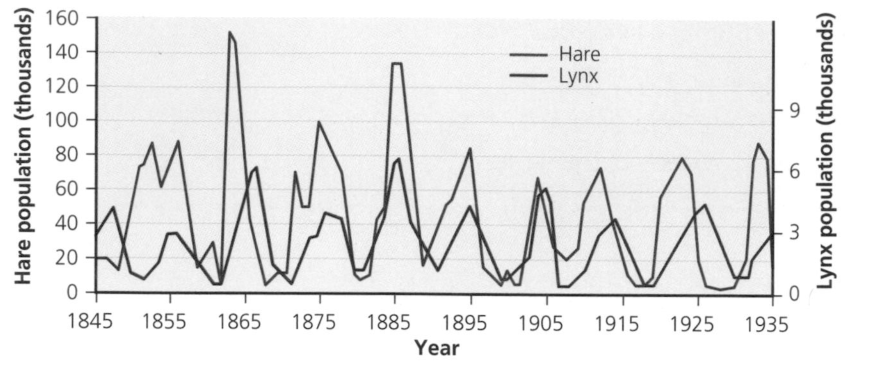

Figure 1.4 Changes in the population size of the snowshoe hare and the lynx between 1845 and 1935

1. A large population of prey provides a lot of food, so the predators can thrive and reproduce in large numbers.
2. The increased predator population results in a decrease in the prey population.
3. The smaller prey population means that the predators can raise fewer young.
4. And so it goes on. The population of predators is always at a lower level than the prey and the changes lag behind the prey.

Examiner's tip

The snowshoe hare and the lynx example will probably not be used in the exam because it is too well known. However, the same basic principles apply to any predator–prey relationship.

Human populations

Population size and structure

Revised

As countries become more developed, there will be a change in the birth rate and death rate that will affect the population size and structure. This is called demographic transition and it has four key stages (Figure 1.5).

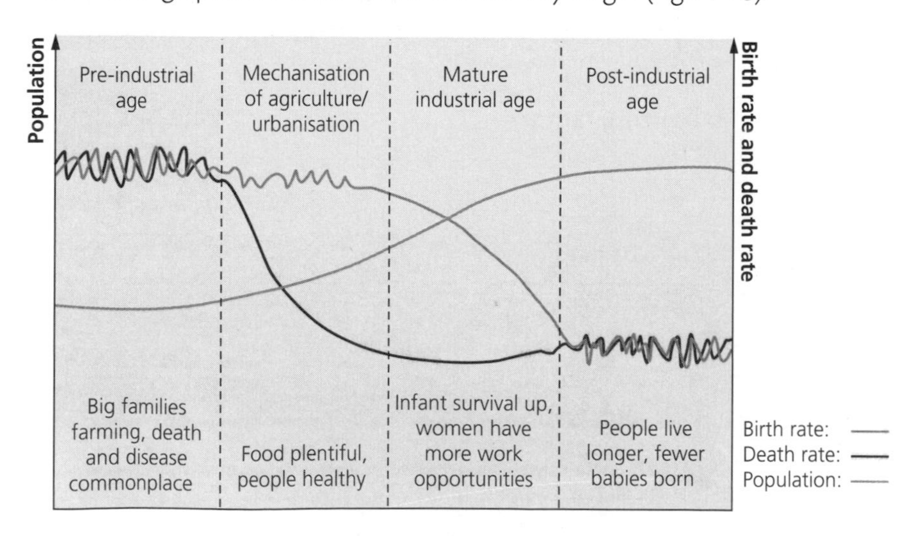

Figure 1.5 The demographic transition model

Stage 1. Initially, birth rate and death rate are very high. There is little education, healthcare and contraception. Many children die early from disease and malnutrition. The average **life expectancy** is low.

Stage 2. Death rate slows down due to better healthcare such as vaccination and access to clean water. Birth rate remains high, so there is rapid population growth.

Stage 3. Birth rate falls. Education is better, especially for women, who have more work opportunities and access to contraception.

Stage 4. Low death and birth rates. Population stabilises. Most developed countries are in this final stage and some are actually in decline, suggesting a fifth stage.

Now test yourself Tested

5 In 1900, the average life expectancy of Brazilians was 54. By 2000 it had increased to 67. Suggest three reasons for this increase.

Answer on p. 110

Population growth rate

Revised

In any period of time, it is obvious that if there are more births than deaths the population will increase. Immigration and emigration are also important, giving us this formula for **population growth rate**:

population growth = (births + immigration) − (deaths + emigration)

In order to compare countries or populations of different size, **percentage growth rate** can be used:

$$\text{percentage growth rate} = \frac{\text{population change during period}}{\text{population at start of period}} \times 100$$

Birth rate and **death rate** are usually given per 1000 of population. However, when we talk about particular causes of death or disease, if numbers are relatively low we can also give the figures per 10 000 or per 100 000.

$$\text{birth rate} = \frac{\text{number of births per year}}{\text{total population that year}} \times 1000$$

$$\text{death rate} = \frac{\text{number of deaths per year}}{\text{total population that year}} \times 1000$$

Now test yourself

6 In 2007, the population of Zimbabwe was 10 million. In 2011 it was 12 million. Calculate its average percentage growth rate per year.

Answer on p. 110

Tested

Age-population pyramids

Revised

You can tell a lot about a country and its stage of demographic transition by looking at its **age-population pyramid**. Figure 1.6 shows age-population pyramids for France (a developed country) and India (a developing country). France is at stage 4 of demographic transition because its birth rate is low and falling while many people are surviving into old age. India is at stage 2: mortality is higher in each age group and life expectancy is lower, but the birth rate is very high and so the population is increasing.

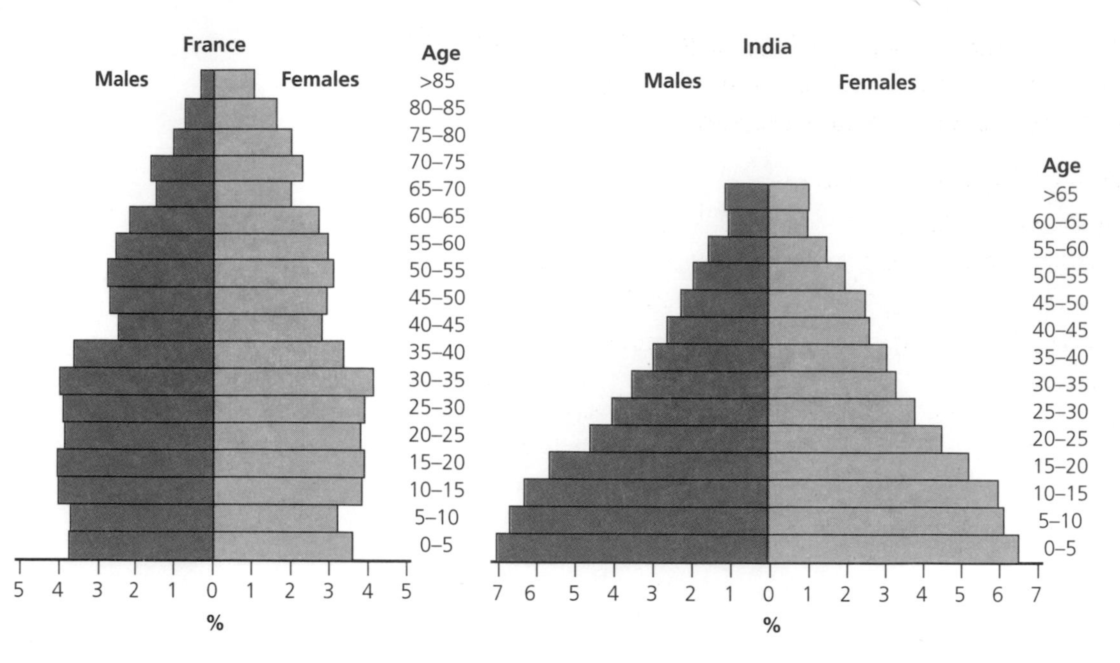

Figure 1.6 Age-population pyramids for France and India

Survival rates and life expectancy

Revised

The percentage of a population surviving at each age can be used to make a **survival curve** (Figure 1.7). Average life expectancy can be calculated from the graph: it is the age at which 50% of the population are still alive, so align your ruler at the 50% survival value and read across.

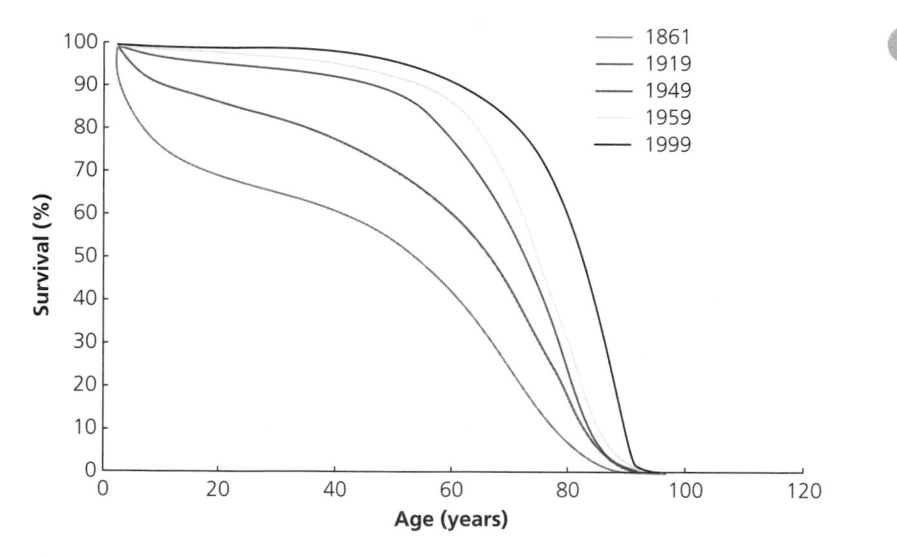

Figure 1.7 A survival curve

Now test yourself

7 Study Figure 1.7. What was the average life expectancy in these years?
  (a) 1861
  (b) 1949
  (c) 1999

8 Suggest reasons for the:
  (a) average life expectancy in 1861
  (b) difference in 1949
  (c) difference in 1999

Answers on p. 110

Tested

An introduction to statistics

Null hypothesis and experimental design

Revised

In science we are interested in explanations, i.e. relating cause to effect. What causes lung cancer? Why does moss grow only on one side of a tree? So we make observations and come up with ideas that we can test. Progress is made by thinking up hypothesis, gathering data, analysing it and coming to a conclusion such as 'No, it couldn't' or 'Possibly, it could'. You can never prove anything, but you can fail to disprove a hypothesis and in doing so you gather support for a particular idea.

A hypothesis is a testable idea. There are two types:

- The **experimental hypothesis** states that there will be a significant connection between cause and effect. An example would be 'Breast-fed babies have fewer allergies in later life then bottle-fed ones.'
- The **null hypothesis** takes a negative view, i.e. 'There is no correlation between bottle feeding and allergies in later life.' A key point here is that statistical tests only test a null hypothesis. They are needed to tell you whether your results are due to chance or not.

Statistical tests

Revised

You need to know how to use three statistical tests:

- standard error and 95% confidence limits
- Spearman rank correlation
- chi-squared (χ^2)

Standard error and 95% confidence limits

Standard error and 95% confidence limits is statistical test used to find out if there is a significant difference between two means. It involves measuring the same variable from two different samples and builds on the **standard deviation** that was covered in Unit 2.

Examiner's tip

It is always better to say *mean* rather than *average*.

Worked example

A student carried out an investigation into the effect of temperature in the growth of duckweed, a tiny, fast-growing aquatic plant that can be grown on petri dishes. She put 20 duckweed plants into 60 shallow dishes of water and kept half at 12°C and half at 22°C. Six days later she counted the number of plants in each dish (Table 1.1).

Table 1.1 Number of duckweed plants in each dish after six days

Dishes kept at 12°C					
37	38	38	36	36	43
40	32	37	39	38	36
40	38	38	39	34	39
39	38	34	37	38	37
35	39	38	41	37	39

Dishes kept at 22°C					
60	56	54	60	58	62
62	61	60	63	59	53
58	57	59	58	61	57
61	59	61	60	56	62
64	53	58	61	60	59

Now test yourself

9 Suggest a null hypothesis for this investigation.

10 Suggest three variables that would need to be controlled.

Answers on p. 110

She used her calculator to work out the mean and standard deviation of each of the samples (Table 1.2).

Table 1.2 Mean and standard deviation for the duckweed plants

	Dishes at 12°C	Dishes at 22°C
Mean	37.7	59.1
Standard deviation (*SD*)	2.2	2.7

She then calculated the standard error of the mean, *SE*, for each sample using the following formula:

$$SE = \frac{SD}{\sqrt{n}}$$

where *SD* is standard deviation and *n* is sample size.

Table 1.3 shows these results. The 95% confidence limits are 1.96 standard errors above the mean and 1.96 standard errors below. This is usually rounded up to 2. If these two ranges do not overlap, we can be 95% confident that the two means are different. In other words, there is a significant difference between the means at the $p = 0.05$ level of probability.

Table 1.3 Results for the duckweed plants

	Dishes at 12°C	Dishes at 22°C
Mean	37.7	59.1
Standard deviation (*SD*)	2.2	2.7
Standard error (*SE*)	0.4	0.5
2 × standard error	0.8	1.0
95% confidence limits	36.9 to 38.5	58.1 to 60.1

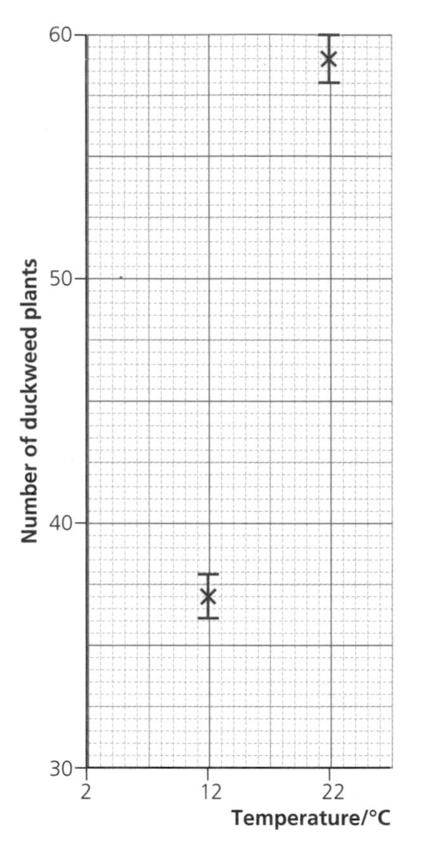

Figure 1.8 How to interpret standard error

Figure 1.8 shows this information plotted on a graph. The mean values for each sample have been plotted as a cross and the bars above and below the crosses represent 2 standard errors on either side of these mean values. In this case, there is no overlap between the bars. Therefore, the student can say that there is a **probability** of less than 0.05 or 5% that these results occurred by chance. She is 95% confident that temperature is having an effect on the growth of duckweed.

Spearman rank correlation

Spearman rank correlation is a test to determine the degree of correlation between two sets of data. Imagine you are using the hypothesis: 'The lighter the seed, the further it drops from the tree.'

In this case, you would get two sets of ranked data:

- the weight of the seeds, from highest to lowest
- the distance the seeds fell from the tree, from highest to lowest

The test can be used for between 7 and 30 pairs of data. You then need to calculate the correlation coefficient, R_s. If the data are positively correlated, the lightest seeds will usually go furthest.

If you apply the Spearman rank correlation test to the data:

- a perfect correlation would give a score of +1, which would show that the lightest seeds dropped furthest from the tree
- a perfect negative correlation would give a score of −1, which would show that the heaviest seeds went the greatest distance

Now test yourself

11 Suggest a null hypothesis for this Spearman rank correlation investigation.

Answer on p. 110

Tested

The R_s value that is significant depends on the number of pairs of data. With five pairs, the significant value is 1. With more pairs, the significant value is lower. If you are given a Spearman rank calculation in Unit 6, you will be given all the values you need.

Chi-squared

Chi-squared (χ^2) is used to decide whether observed results are significantly different from those that you would expect if they were due to chance. Like most statistical tests, the chi-squared test gives you a result in terms of probability, expressed as a value of *p*. If the value of *p* is less than 0.05 (meaning 5% probability), you can say the results are significant and so reject the null hypothesis. It means that you are 95% certain that the results are not just random.

The chi-squared test is suitable if you can answer 'Yes' to these questions:

- Does the data fall into particular categories?
- Is there an expected outcome? This is where the null hypothesis has a direct use. There will be no significant difference in the categories if the null hypothesis is correct. The values in all categories will be similar.

Worked example

Do birds have a preferred habitat for feeding?

The numbers of five species of bird were observed feeding over a period of three days. The results are shown in Table 1.4.

Table 1.4 Numbers of birds feeding

Species	Total number of birds observed		
	Woodland	**Open farmland**	**Gardens**
Great tit	12	13	12
Long-tailed tit	2	1	0
House sparrow	0	0	13
Blue tit	53	43	30
Greenfinch	6	5	22

Looking at the results, common sense should tell you that, based on this data:

- great tits have no feeding preference
- house sparrows have a definite feeding preference
- there is not enough data on long-tailed tits to draw any conclusions at all

But what about blue tits and greenfinches? Are these results significant or not? This is where chi-squared can be useful. The formula for the test is:

$$\chi^2 = \Sigma \frac{(O - E)^2}{E}$$

where *O* is the observed result and *E* is the result we would expect if there was no significant difference.

In this example we will use the data on blue tits.

Step 1. The experimental hypothesis will be something like 'Blue tits have a preference for the type habitat in which they feed.' There are lots of other ways of wording this. The null hypothesis is 'Blue tits have no preference for the type of habitat in which they feed.'

Examiner's tip

With the chi-squared test, the null hypothesis is always along the lines of 'There is no difference between the observed results and the expected results.'

Examiner's tip

The Unit 4 exam will not contain a full statistical calculation like this — you will only get one in the Unit 6 practical exam — but it might contain questions that ask you for a null hypothesis or to interpret the value of *p* using the words *probability, significant* and *chance.*

Step 2. Work out the values of the observed results and the expected results. The results in the table give us the observed results (*O*). They are 53, 43 and 30. We have to calculate the expected results (*E*). If the feeding behaviour of blue tits was purely a matter of chance, we would expect equal numbers to visit each habitat. So the expected number visiting each habitat is the mean, calculated as:

$$53 + 43 + 30 = 126$$

$$\frac{126}{3} = 42$$

In other words, if there were no preference, you would expect 42 visits to each type of habitat.

Step 3. Calculate χ^2. We now use the formula to calculate the value of χ^2. The best way to approach it is to write the values in a table such as Table 1.5, so that each row is the next step in the calculation.

Table 1.5 Calculating χ^2

	Woodland	Open farmland	Gardens
Observed results (*O*)	53	43	30
Expected results (*E*)	42	42	42
$(O - E)^2$	121	1	144
$\frac{(O - E)^2}{E}$	2.88	0.02	3.43

In the formula, the symbol Σ means 'the sum of all', so we have to add together all the separate values of the last line in the table. In this example, our χ^2 value is:

$$2.88 + 0.02 + 3.43 = 6.33$$

Step 4. Look up and interpret the values of χ^2. We now need to look at Table 1.6, which is a table of values of χ^2. We must work out the number of **degrees of freedom** in order to look in the right row of the table. The number of degrees of freedom is one less than the number of categories. We have three categories, so the number of degrees of freedom is 2.

We now have a χ^2 value of 6.33 and a number of degrees of freedom of 2. We can use these figures to find out if our χ^2 value is above or below the critical 5% value. If the probability is less than 5%, we can say that our results are significant because there is a low probability that they are due to chance.

Table 1.6 Chi-squared values

Degrees of freedom	Probability					
	0.50	0.25	0.10	0.05	0.02	0.01
1	0.45	1.32	2.71	3.84	5.41	6.64
2	1.39	2.77	4.61	5.99	7.82	9.21
3	2.37	4.11	6.25	7.82	9.84	11.34
4	3.36	5.39	7.78	9.49	11.67	13.28

We have 2 degrees of freedom, so the appropriate row is highlighted. Our chi-squared value of 6.33 falls between 5.99 and 7.82, so the probability that our results are due to chance is between 0.05 and 0.02 (i.e. between 5% and 2%). The value is below 5%, so we can say the results are significant. We can reject the null hypothesis and state that blue tits *do* have a preference for the habitat in which they feed.

Exam practice

1 In an investigation, the population density of plants in a regularly cut lawn was compared with that in a lawn which was cut only occasionally. The results are shown in the table below.

Species	Mean population density/number of plants per m²		Result of statistical test
	Regularly mown lawn	Occasionally mown lawn	Value of *p*
Clover	35.0	18.1	< 0.02
Dandelion	10.8	3.4	< 0.05
Ragwort	1.2	8.7	< 0.01
Buttercup	1.3	10.1	< 0.01
Ribwort plantain	4.3	2.8	> 0.50

(a) Describe how you could use quadrats to find the mean population density of ragwort plants in a lawn. [4]

(b) Give the null hypothesis for this investigation. [1]

(c) Use the words *probability* and *chance* to explain what is meant by $p < 0.05$. [2]

(d) What conclusions can be drawn from the results of this investigation? [3]

2 In April greenfly hatches from eggs that were laid on a rose bush the previous year. The greenfly begin to reproduce asexually. All the offspring are wingless females, born with young greenfly already developing inside them. After a few months there is a large population on the plant. At this point the aphids begin to give birth to winged males and females that can fly away, mate and lay eggs on new bushes.

(a) Suggest why it is an advantage to reproduce asexually in spring. [3]

(b) Competition is a *density-dependent factor*. Explain what this term means. [1]

(c) List two limiting factors that could slow down population growth. [1]

(d) Suggest the advantages of the following:

(i) wings [1]

(ii) sexual reproduction [1]

(iii) laying eggs [1]

3 A student wanted sample the diversity of plant species in a meadow. She needed know how many quadrats to use in the investigation, so she added up the number of species identified. Explain how she could use the following graph to decide how many quadrats to use. Use point X to help you. [2]

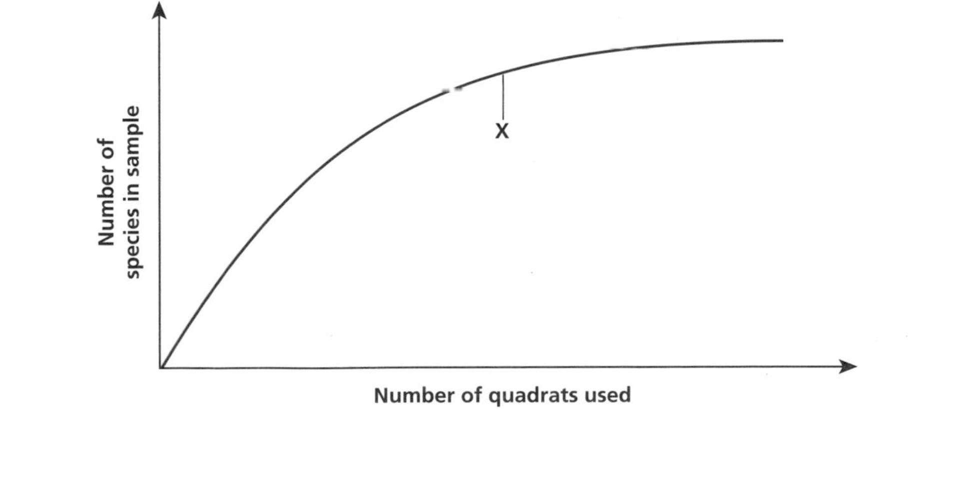

Answers and quick quizzes online

Online

Examiner's summary

By the end of this chapter you should be able to understand:

- ✔ In a habitat, a population consists of all the individuals of one species, whereas a community consists of all the individuals of all the species.
- ✔ Each species occupies a particular niche, governed by its adaptation to both biotic and abiotic conditions.
- ✔ Quadrats and transects are used for sampling data. Quadrats compare different areas of land, whereas transects show gradual change from one area to another.
- ✔ Animal population size can be estimated using the mark-release-recapture method.
- ✔ Population size varies according to the effect of abiotic factors and interactions between organisms such as interspecific and intraspecific competition and predation.
- ✔ How to interpret growth curves, survival curves and age-population pyramids.
- ✔ How to calculate population growth rates from data on birth and death rates.
- ✔ How to relate changes in the size and structure of human populations to different stages in demographic transition.

2 ATP, photosynthesis and respiration

ATP

Synthesis of ATP

Revised

Adenosine triphosphate (ATP) is a substance found in all organisms and its function is to deliver instant energy in usable amounts. All organisms respire all the time because they need a constant supply of ATP. It is a relatively simple molecule that releases its energy by splitting into **adenosine diphosphate (ADP)** and **phosphate**, which can be written as PO_4^- or P_i which stands for inorganic phosphate.

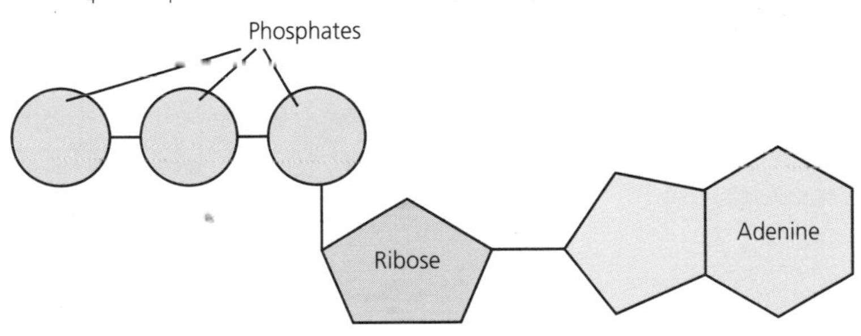

Figure 2.1 An ATP molecule

Examiner's tip

You don't need to memorise the detailed structure of ATP, just to understand how it provides energy.

ATP:

- releases instant energy because it needs just one simple chemical step
- releases energy in usable amounts — if it released more energy than needed, the excess would be wasted as heat
- is a relatively small molecule so it can diffuse rapidly around the cell
- is an unstable molecule that cannot be stored, so it must be constantly resynthesised

There is only a certain amount of ATP in a cell. It is constantly being broken down and needs to be resynthesised by respiration. In humans and all other warm-blooded animals that respire quickly, the weight of ATP produced each day is greater than the entire body weight. We have a relatively small amount of ATP, but it is being constantly broken down and remade.

The balance of ATP and ADP/P_i in a cell is a bit like a battery. If it is all ATP, the battery is fully charged. If it is mostly ADP and P_i the battery is run down and needs to be recharged by the process of respiration.

There are many processes that use ATP, but for exam purposes the three main ones are:

- muscular contraction
- active transport
- protein synthesis

Typical mistake

Students often lose marks by writing P for phosphate. However, P is the symbol for the element phosphorous, not phosphate. You can use the word phosphate or the abbreviations P_i or PO_4^-

Oxidation and reduction

Revised

Before you can revise respiration and photosynthesis, you need to understand about oxidation, reduction and **coenzymes**. Photosynthesis and respiration are both energy-transfer processes. The energy is transferred in electrons:

- if a substance loses electrons, it loses energy and has been **oxidised**
- if a substance gains electrons, it gains energy and has been **reduced**

When one substance is oxidised another is usually reduced, so they are collectively known as **redox** reactions.

A key element in both respiration and photosynthesis is the need to carry electrons and this job is done by coenzymes. The coenzyme in respiration is **NAD$^+$**. When it picks up an electron, it becomes **NADH** or **reduced coenzyme**. The coenzyme in photosynthesis is **NADP$^+$**, which becomes **NADPH** or, again, just reduced coenzyme.

Coenzymes are complex organic molecules that are used to transfer the products from one reaction to become the substrate in another reaction.

Examiner's tip

Whenever you see NADH or NADPH, just think 'an electron being carried'.

Examiner's tip

NADH is the reduced coenzyme in respiration, whereas NADPH is the reduced coenzyme in photosynthesis — think P for photosynthesis. The P doesn't actually stand for photosynthesis, but who cares?

Photosynthesis and respiration as opposites

Revised

The equation for respiration is:

glucose + oxygen → carbon dioxide + water + ATP

The equation for photosynthesis is:

carbon dioxide + water + sunlight → glucose + oxygen

Therefore, in respiration glucose is oxidised, producing carbon dioxide and releasing energy. In photosynthesis carbon dioxide is reduced using sunlight energy and producing glucose.

Photosynthesis

Key concepts

Revised

Photosynthesis is vital to life on Earth because it:

- is the only way that energy can get into ecosystems
- turns a colourless, odourless gas (carbon dioxide) into solid organic molecules such as glucose and starch, i.e. it makes food
- creates oxygen as a by-product

Chloroplasts are the organelles of photosynthesis (Figure 2.2). They give the green colour to all upper parts of a plant, but they are mainly concentrated in the **palisade cells** of the leaves. Chloroplasts were studied in Unit 2. The key features are:

- a light-harvesting group of compounds collectively called **chlorophyll**
- the chlorophyll is embedded in flat discs called **thylakoids**

- the thylakoids are packed into stacks called **grana** (singular: granum)
- between the grana is an enzyme-rich fluid called the **stroma**

Stroma is the fluid inside a chloroplast.

Photosynthesis is split into two key steps: the light-dependent reaction and light-independent reaction.

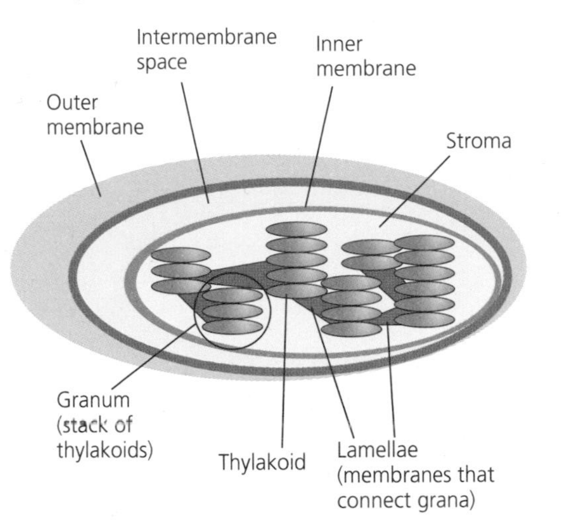

Figure 2.2 The structure of a chloroplast

Measuring the rate of photosynthesis

Revised

Photosynthesis is usually measured using an aquatic plant, often *Elodea* or *Cabomba*. This is because the oxygen is given off as bubbles that can be collected easily (Figure 2.3).

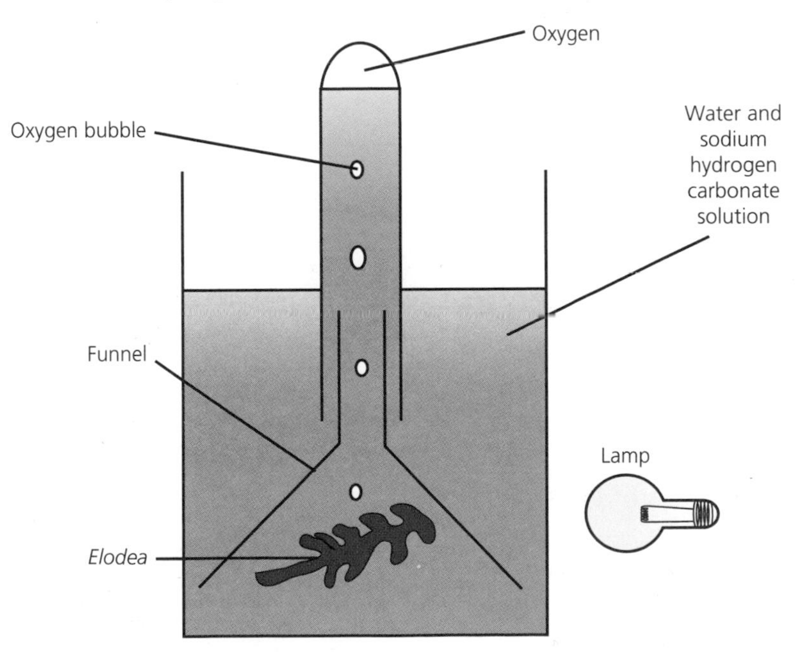

Figure 2.3 Measuring photosynthesis

Using this apparatus you can investigate the effect of various limiting factors on the rate of photosynthesis:

- Aquatic plants get their carbon dioxide in solution, as HCO_3^- ions. Therefore, carbon dioxide levels can be varied by using different concentrations of HCO_3^-.
- Light intensity can be varied by placing a lamp at different distances from the plant. Light intensity is proportional to $1/d^2$ where

d = distance. This means that if you double the distance, the light intensity will be one-quarter of the original value.

- Temperature can be varied by heating or cooling the water.

Light-dependent reaction

The process

Revised

This first process relies on chlorophyll, so it takes place on the thylakoids. The key steps are as follows:

- Light hits chlorophyll, which reacts by emitting two **excited** electrons.
- The electrons pass along an electron transport chain in the thylakoid membranes, where the energy is used to make two vital compounds: ATP and reduced NADP (also called NADPH). They possess the energy and the reducing power needed to turn carbon dioxide into glucose.
- In order for photosynthesis to continue, chlorophyll needs replacement electrons. These come from splitting water in a process called **photolysis**.
- Photolysis creates electrons, hydrogen ions and oxygen gas as a by-product.

Excited in this context means raised to a higher energy level.

Examiner's tip

When studying this topic, stick to the specification. You don't need anything more than the details here. If you come across photosystem 1 and photosystem 2, you are going into too much detail.

Now test yourself

Tested

1. Apart from light, list four things that are needed for the light-dependent reaction.
2. In the light-dependent reaction, name the two compounds that are necessary in order for the plant to reduce carbon dioxide into glucose.
3. Explain how photosynthesis produces oxygen.

Answers on p. 110

Light-independent reaction

The process

Revised

The light-independent reaction takes place in the stroma of the chloroplasts and involves the reduction of carbon dioxide into glucose. It is sometimes called the **Calvin cycle** (Figure 2.4). The key steps are as follows:

- One molecule of carbon dioxide combines with a 5-carbon compound called **ribulose bisphosphate (RuBP)**.
- This results in a 6-carbon compound that immediately splits into two molecules of **glycerate 3-phosphate (GP)**. GP has three carbon atoms but is not a sugar.
- ATP and reduced NADP (from the light-dependent reaction) bring about the reduction of GP to **triose phosphate (TP)**, which is the first sugar.
- Triose phosphate is eventually turned into **glucose**.

- Every cycle of reactions increases the number of carbon atoms by one — the one in carbon dioxide. So it takes six cycles to accumulate enough carbon to make one molecule of glucose. Each cycle also regenerates RuBP so that the cycle can continue.

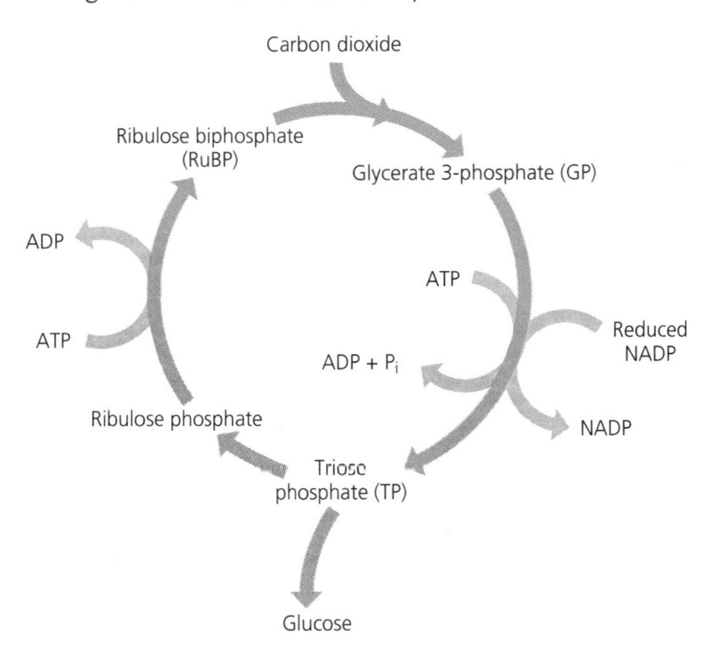

Figure 2.4 A simple Calvin cycle

Examiner's tip

The specification refers to photosynthesis in a typical C3 plant, which means that the first sugar made (triose phosphate) has three carbons. Most plants are C3 plants; there are C4 plants, but you don't need to know about them.

Examiner's tip

The specification states that 'Triose phosphate is converted to useful organic substances', which in plain English means glucose, starch, sucrose and various other organic molecules. Plants also synthesise proteins, phospholipids and nucleic acids using the mineral ions — notably nitrate and phosphate — absorbed by the roots.

Revision activity

Make sure you can draw a diagram to show the light-independent reaction, showing the three intermediates (RuBP, GP and TP) and the role of ATP and reduced NADP.

Limiting factors

What limits the rate of photosynthesis?

A limiting factor is one which, if the supply is increased, will speed up the process. The limiting factors in photosynthesis are:

- **temperature** — the light-independent reaction is temperature-sensitive because it is enzyme-controlled. The light-dependent reaction is less so
- **carbon dioxide** — needed as the source of carbon
- **light intensity** — needed to excite the chlorophyll

Knowing about limiting factors is important if you want to make plants grow faster — a basic requirement of agriculture.

Generally, increasing the amount of light will increase the rate of photosynthesis until carbon dioxide becomes limiting. Then, increasing the amount of carbon dioxide will increase the rate until temperature becomes limiting. When the temperature is at an optimum the plant will be photosynthesising at its maximum rate, which is then just limited by the amount of chlorophyll it possesses.

For maximum growth, plants also need a supply of **mineral ions** — the key ones are **nitrate**, **phosphate** and **potassium**. That is why fertiliser is needed (see p. 36).

Respiration

ATP synthesis

Revised

Respiration is one of the seven signs of life. Its purpose is to transfer the energy from organic molecules, such as glucose and lipid, into ATP. There are two types of respiration:

- **aerobic respiration**, which requires oxygen
- **anaerobic respiration**, which does not require oxygen

Aerobic respiration has four stages: **glycolysis**, which takes place in the cytoplasm, followed by the **link reaction**, **Krebs cycle** and **electron transport chain**, all of which take place in the **mitochondria**.

Anaerobic respiration is basically glycolysis that cannot go any further, usually because there is no oxygen.

Examiner's tip

Don't say that 'respiration *makes* energy' because it doesn't. Respiration *releases* the energy stored in organic molecules and *transfers* it into ATP. 'Transfer' or 'release' are good words. 'Makes' or 'creates' are crimes against physics and you will have broken the first law of thermodynamics.

Mitochondria

Revised

Mitochonrdria are the organelles of aerobic respiration. They are found in eukaryotic cells, where they are the source of most of the cells' ATP. A mitochondrion is bounded by two **membranes**. The inner membrane is the site of the electron transport chain and membrane-bound ATP synthase enzymes. It is folded into cristae, which increases the surface area of the membrane and allows increased electron transport and ATP synthesis.

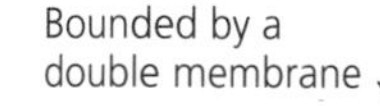

Figure 2.5 A mitochondrion. Full aerobic respiration consists of four processes, the last three of which take place in the mitochondria. The link reaction and the Krebs cycle take place in the fluid matrix in the centre, whereas the electron transport chain takes place on the inner membrane

Glycolysis

Revised

Glycolysis means 'sugar splitting' and it is a universal process — it happens in the cytoplasm of all cells. Aerobic and anaerobic respiration both start with glycolysis. The key features are:

- glucose is raised to a higher energy level using two molecules of ATP

- the resulting molecule is split into two molecules of **pyruvate**, a 3-carbon compound
- the splitting of pyruvate creates two vital compounds: four molecules of ATP, giving a net profit of two, and two molecules of reduced coenzyme NADH

What happens next — either aerobic or anaerobic respiration — depends on whether or not oxygen is available.

Aerobic respiration

The link reaction

Revised

If oxygen is available, and it is a eukaryotic cell, pyruvate passes into the mitochondria where it enters the link reaction, also called **pyruvate oxidation**. In the matrix of the mitochondria, pyruvate is split to form acetate (a 2-carbon compound) and carbon dioxide. The acetate combines with **coenzyme A** to produce **acetylcoenzyme A** (sometimes just called acetyl coA). This process produces no ATP but it does make NADH.

Krebs cycle

Revised

The Krebs cycle is a series of reactions the purpose of which is to remove electrons (and hence energy) from what is left of the glucose, now just acetate. It begins when acetyl coA combines with a 4-carbon compound to become a 6-carbon compound (Figure 2.6).

Examiner's tip

You don't need to know the names of the compounds involved in the Krebs cycle.

The 6-carbon compound enters a series of reactions in the matrix, which produces:

- carbon dioxide
- ATP
- reduced coenzymes NADH and $FADH_2$
- more 4-carbon compounds to continue the cycle

$FADH_2$ is similar to NADH and does much the same electron-carrying function.

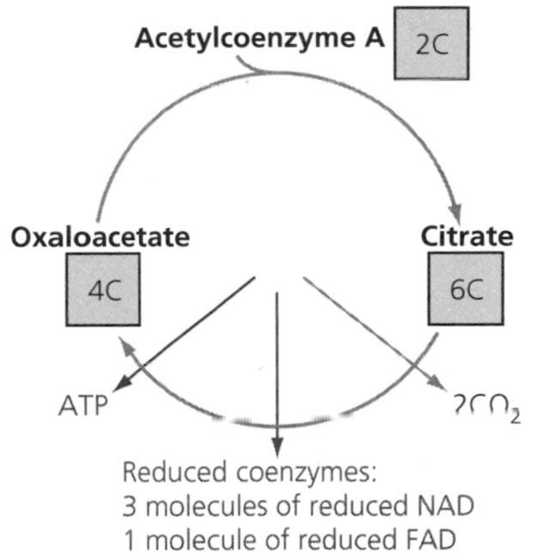

Acetylcoenzyme A 2C
Oxaloacetate 4C
Citrate 6C
ATP
$2CO_2$
Reduced coenzymes:
3 molecules of reduced NAD
1 molecule of reduced FAD

Figure 2.6 The essentials of the Krebs cycle. The cycle turns twice for each glucose molecule

The electron transport chain

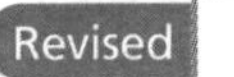

This process makes a lot of ATP by the process of **oxidative phosphorylation** and takes place on the inner mitochondrial membrane. It involves all the reduced coenzymes made by the first three processes and uses the energy in the electrons to make ATP:

Oxidative phosphorylation is that part of the respiratory pathway in which energy released in the electron transfer chain is used in the production of ATP.

- Reduced coenzymes deliver their electrons to the proteins of the electron transport chain.
- The electrons pass along a the electron transport proteins, which are first reduced and then oxidised. These redox reactions release energy that is used to pump **protons** (or H^+ ions) into the outer mitochondrial space. The result is a high concentration of H^+ ions in the outer mitochondrial space.
- The protons diffuse back into the matrix through the middle of ATPase enzymes. As they pass through the enzymes, ATP is synthesised.

Outer membrane of mitochondrion

(1) (2) (3)

Outer mitochondrial space

H^+ H^+ H^+ H^+

Inner membrane of mitochondrion

Reduced NAD → NAD + H^+

$2H^+ + \frac{1}{2}O_2$ → H_2O

ADP + P_i → ATP

ATP synthase

Oxidative phosphorylation

Electrons (e^-)

Protons (hydrogen ions) (H^+)

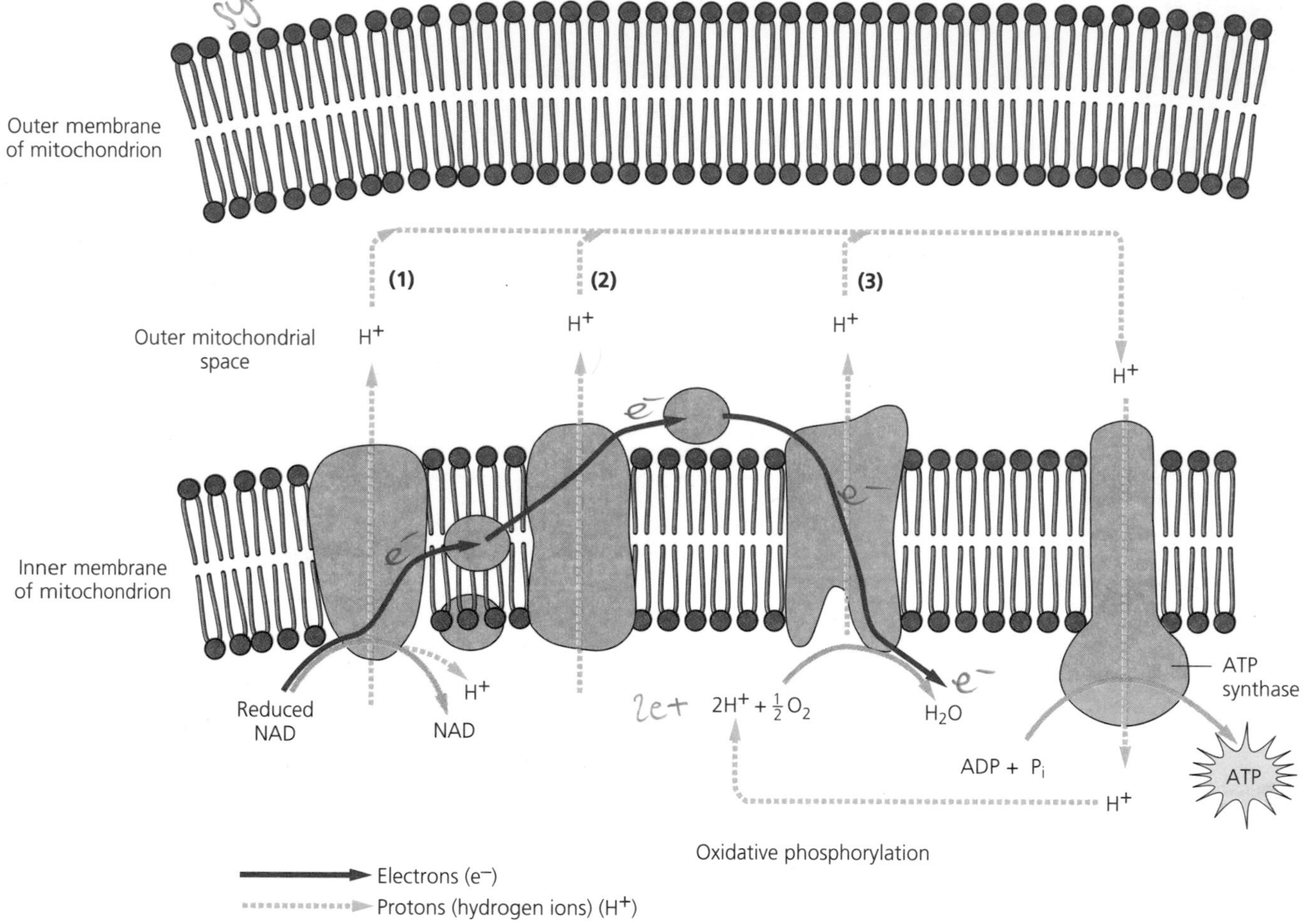

Figure 2.7 The inner mitochondrial membrane. ATP is made by oxidative phosphorylation on the inner mitochondrial membrane

Why is oxygen important?

The end product of the process is low energy electrons that need to be mopped up, which is why oxygen is needed. The oxygen combines with the electrons and hydrogen ions to form water:

$$4H^+ + 4e^- + O_2 \rightarrow 2H_2O$$

If oxygen is in short supply, there is nothing to accept the electrons at the end of the process, so the electron transport chain stops and no ATP is produced. This is fatal for most organisms in a very short period of time — so keep breathing.

RESPIRATION.

Phosphorylation

Revised

There are two types of **phosphorylation**. ATP is made in glycolysis and the Krebs cycle by **substrate-level phosphorylation**, which means that the phosphate used to make ATP comes from a substrate, i.e. another molecule. The other way to make ATP is by **oxidative phosphorylation**, which takes place in the electron transport chain. Here, ATP is made by a series of electron transfers followed by the diffusion of hydrogen ions across the mitochondrial membrane.

Phosphorylation means the addition of a phosphate.

Examiner's tip

If a question says 'Describe how oxidative phosphorylation makes ATP', in plain English it means 'Tell us about the electron transport chain.'

Now test yourself

Tested

4 Of the four processes in aerobic respiration, list all those that make ATP.
5 Explain what is meant by:
  (a) substrate-level phosphorylation
  (b) oxidative phosphorylation

Answers on p. 110

How much ATP is made in aerobic respiration?

Revised

So far, for every glucose molecule, we have:

- 2 ATP molecules from glycolysis, which is all that anaerobic respiration produces
- 2 ATP molecules from the Krebs cycle

and that's it for substrate-level phosphorylation. However, there are many reduced coenzymes carrying electrons that are used to power oxidative phosphorylation:

- 2 NADH molecules from glycolysis
- 2 NADH molecules from the link reaction
- 6 NADH molecules from the Krebs cycle
- 2 $FADH_2$ molecules from the Krebs cycle

When they deliver their electrons to the inner mitochondrial membrane, each NADH produces 3 ATP molecules and each $FADH_2$ produces 2 ATP molecules. Adding it all up, there are 10 NADH providing 30 ATP molecules and 2 $FADH_2$ providing 4 ATP molecules. So the grand total is 38 ATP molecules, 34 of which come from oxidative phosphorylation.

Examiner's tip

Don't worry too much about learning the exact amounts of NADH and ATP at each stage. The vital point is that oxidative phosphorylation, which is the key process at the end of aerobic respiration, produces far more ATP than any other process.

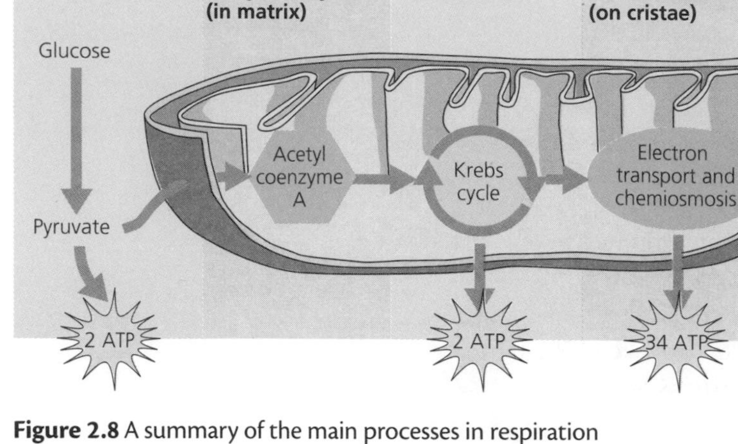

Figure 2.8 A summary of the main processes in respiration

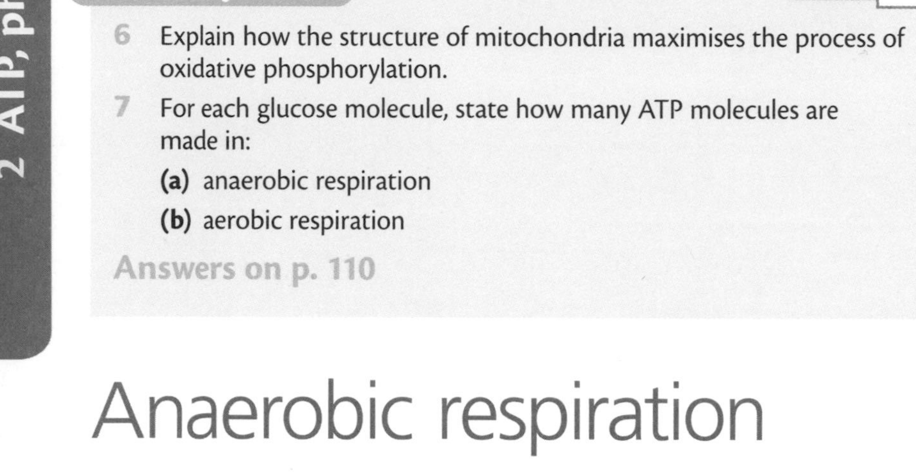

Now test yourself Tested

6 Explain how the structure of mitochondria maximises the process of oxidative phosphorylation.

7 For each glucose molecule, state how many ATP molecules are made in:

(a) anaerobic respiration

(b) aerobic respiration

Answers on p. 110

Anaerobic respiration

What is anaerobic respiration?

Revised

In the absence of oxygen, respiration cannot continue because pyruvate cannot be oxidised further. The cell or organism must survive on the small amount of ATP made in glycolysis. The problem is that glycolysis cannot continue when all of the coenzyme has been reduced to produce NADH. The key step in anaerobic respiration is to reduce the pyruvate so that the coenzyme **NAD^+ is resynthesised**:

- In animals and many bacteria, the pyruvate is converted into **lactate**.
- In plants and many fungi (including yeast), the pyruvate is converted into **ethanol** (alcohol) and carbon dioxide — the basis of alcoholic fermentation.

Now test yourself Tested

8 Is NAD^+ reduced or not? Explain your answer.

Answer on p. 110

Measuring respiration

Using a respirometer

Revised

Respiration is usually measured by oxygen uptake using a respirometer like the one in Figure 2.9. The organism respires, taking in oxygen and giving out carbon dioxide. Normally, this would not change the volume in the chamber, but the sodium hydroxide absorbs all the carbon dioxide. So it is just as if the organism is not making carbon dioxide at all. As a result, the volume in the chamber reduces as the organism uses oxygen. You can measure how much oxygen is being used by reading the scale.

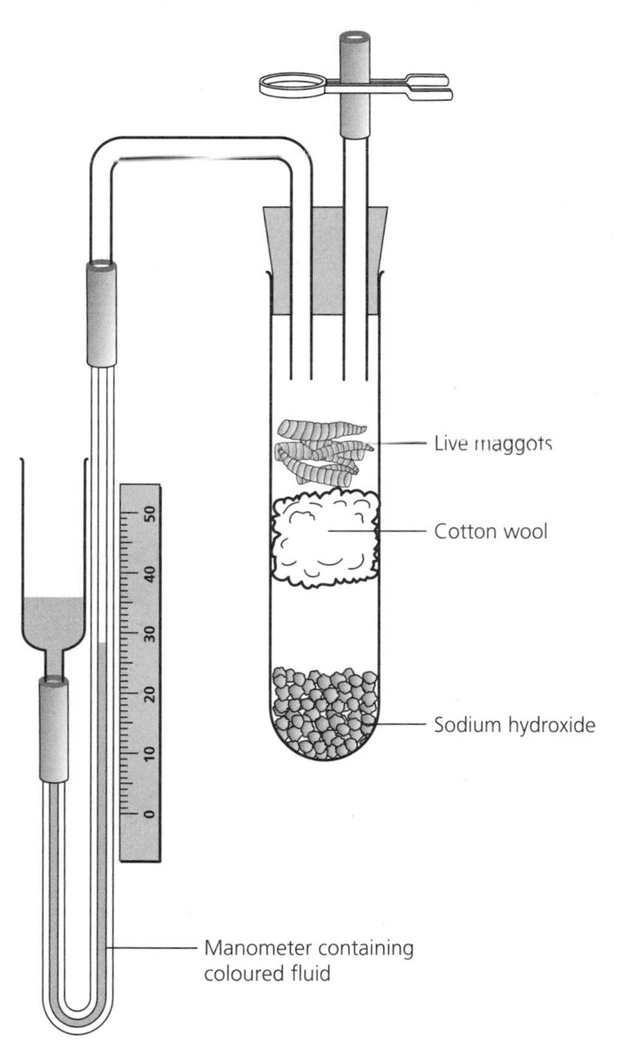

Figure 2.9 A simple respirometer

To compare different organisms, three measurements are needed:

- the mass of the organisms
- the volume of oxygen used
- time taken

Now test yourself

Tested

9 Suggest units for measuring respiration.

Answer on p. 110

Exam practice

1 Complete the following table with a 'yes' or a 'no'. [3]

Question	Photosynthesis	Respiration
Is ATP made?		
Is ATP made in an electron transport chain?		
Is ATP made by substrate-level phosphorylation?		

2 Explain how the structure of the ATP molecule allows it to perform its role in the cell. [4]

Answers and quick quizzes online

Online

Examiner's summary

By the end of this chapter you should be able to understand:

- ATP is made from ADP and phosphate. Its function is to provide an instant source of energy.
- ATP is constantly being used by processes that require energy and resynthesised by respiration.
- Photosynthesis has two key stages: the light-dependent and the light-independent reactions.
- The light-dependent reaction produces ATP and NADPH.
- The light-independent reaction uses ATP and NADPH to convert carbon dioxide into glucose.
- Temperature, carbon dioxide concentration and light intensity are the main limiting factors in the rate of photosynthesis.
- Full aerobic respiration has four stages: glycolysis, the link reaction, Krebs cycle and electron transport chain.
- In glycolysis, glucose is converted into pyruvate.
- In the link reaction, pyruvate is converted into acetylcoenzyme A.
- In the Krebs cycle, the electrons in the acetate are used to make ATP and reduced coenzymes.
- In the electron transport chain, the electrons carried by the reduced coenzymes are used to make ATP by oxidative phosphorylation.

3 Energy transfer and recycling

Energy transfer

Food chains and food webs

Revised

A **food chain** simply shows what eats what, with arrows to show the direction of **energy** flow:

grass → rabbit → fox

This shows that the energy in the grass passes to the rabbit and then to the fox. Food chains are not much use because rabbits eat more than just grass and foxes eat more than just rabbits. **Food webs** such as Figure 3.1 are much more useful in representing the complex interactions that take place in a community.

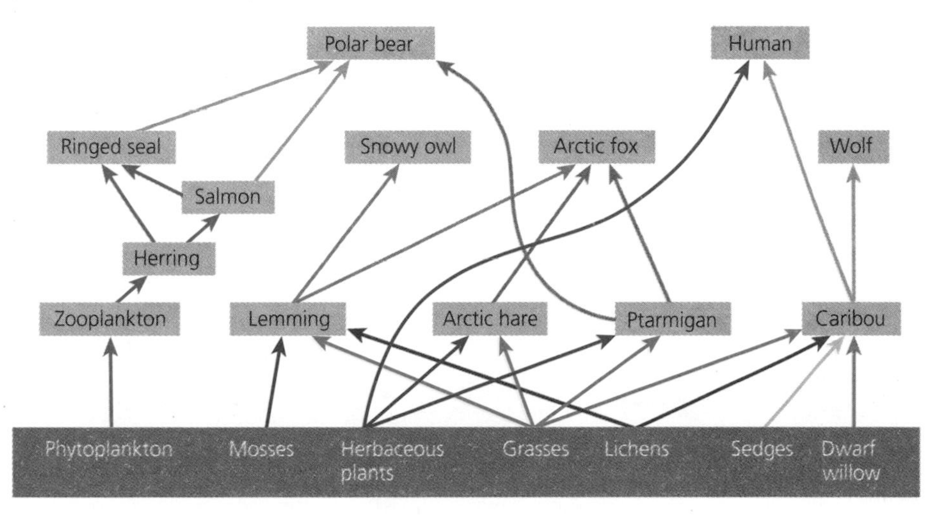

Figure 3.1 A food web of the Arctic tundra. Many food chains interlink to form this complex food web

Exam questions about food webs often ask you to name an organism at a particular **trophic level**. Simply count the arrows from the bottom of the food web to the organism in question. If there is one arrow, it is a primary consumer; two arrows, it is a secondary consumer etc.

Trophic levels are steps in a food chain, such as producers, primary consumers and secondary consumers.

Now test yourself

Tested

1 Name an organism in the food web in Figure 3.1 that is both a tertiary and a quaternary consumer.

Answer on p. 110

Where does all the energy go?

Revised

A vast amount of sunlight energy reaches the Earth, but only a small fraction — perhaps 0.5% — is captured by plants. The rest of this energy is used to:

- heat up the planet as it is absorbed by the atmosphere — this is good because it stops us from freezing
- heat up the oceans and the land — only a small proportion actually makes contact with plants

Of the energy that does reach plants:

- some is reflected off the leaves
- some misses the chloroplasts
- some is the wrong wavelength — most plants absorb red and blue light but reflect green light

Examiner's tip

Energy transfer is a favourite topic in Unit 4 exams. You need to be able to explain why energy transfers are inefficient.

Now test yourself Tested

2 Explain why plants cannot make use of all the energy that reaches the leaves.

Answer on p. 110

Plant material

Revised

A lot of energy is lost when animals eat plants. Imagine the rabbit eats some grass that contains 1000 kJ of energy. A lot of this energy is locked up in cellulose, which is very difficult to digest — animals themselves cannot make the enzyme **cellulase**. Herbivores have microorganisms in their gut that can make cellulase, but even so almost half of the energy in the grass — say 500 kJ — passes straight out of the body and is lost in the faeces. The energy available to the rabbit is that which is contained in the compounds that can be digested and therefore **absorbed** into the blood.

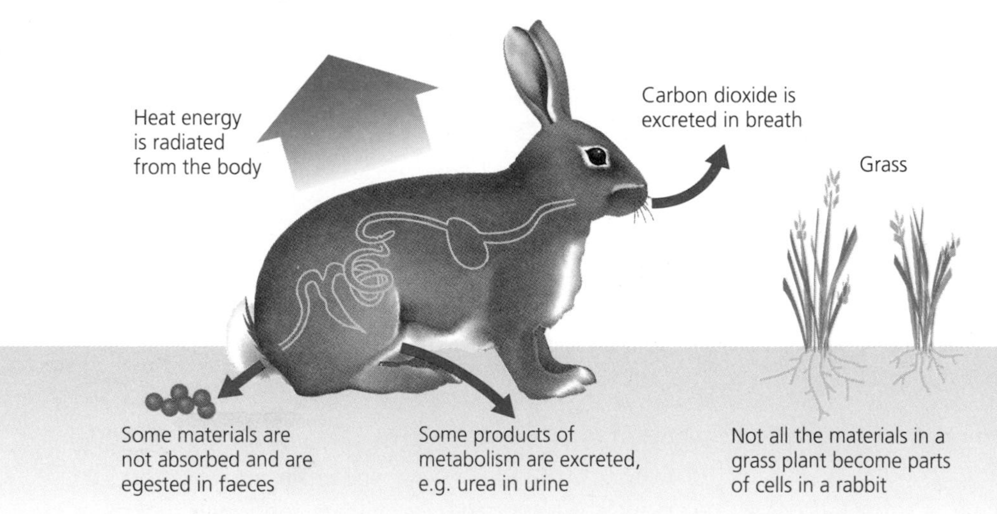

Figure 3.2 Energy transfer through a rabbit

The rabbit, like all mammals, is warm blooded, and maintaining a constant body temperature uses a lot of energy. The vast majority of the energy in the absorbed food is lost to the environment as **heat**, which is released by **respiration**. The figure could be as high as 480 kJ.

Examiner's tip

All organisms lose energy as *heat* which is released during *respiration*. Many candidates fail to appreciate this fact.

Therefore only a small proportion of the energy in the food actually finds its way into the tissues of the animal — what we think of as 'meat' — but this is all that is available to the rest of the ecosystem. In this case, just 20 kJ of the original 1000 kJ — just 2% — is converted into living tissue.

The next animal in the food chain — the carnivore or omnivore — has a slightly easier job because a much higher proportion of animal tissue is digestible. The problem is that it takes much more energy to hunt and catch the food in the first place.

Now test yourself Tested

3 Explain why the energy transfer from plant to herbivore is less efficient than from herbivore to carnivore.

4 Explain why herbivores such as antelope and gazelle spend all day feeding, whereas carnivores such as lions spend most of the day resting.

5 Explain why a snake may only have to feed every few months.

Answers on p. 110

Pyramid diagrams

Revised

The numbers of organisms in a food chain, or their **biomass**, can be shown by simple diagrams. Even when the **pyramids of numbers** are not true pyramids — usually because the producers are large or there are parasites involved — the pyramids of biomass will be as shown in Figure 3.3.

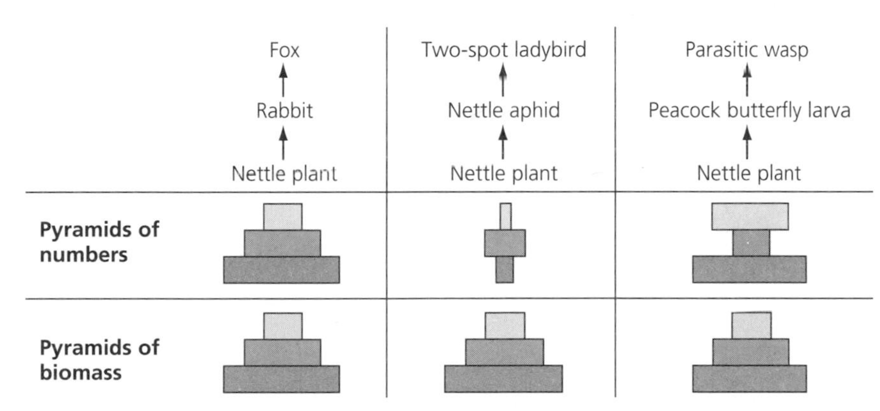

Figure 3.3 Pyramids of numbers and biomass using food chains based on nettles

Pyramids of energy must be pyramids because there cannot possibly be more energy in one trophic level than in the one below. This is basic thermodynamics: all energy transfers are inefficient and some is always lost as heat. Energy transfers in food chains are rarely more than 10% efficient, for the reasons outlined on page 34.

Figure 3.4 Most of the available energy is lost at each trophic level

Biomass simply means the mass of living material, such as the mass of all the grass or all the rabbits. With plants, the water content varies greatly, so **dry mass** would usually be used.

We can see from the pyramids of numbers that the animals at the higher trophic levels get progressively rarer. The reason is simple: the energy runs out. There is not enough energy to support a large population of apex predators. Most food chains have two or three links and there are rarely more than four or five.

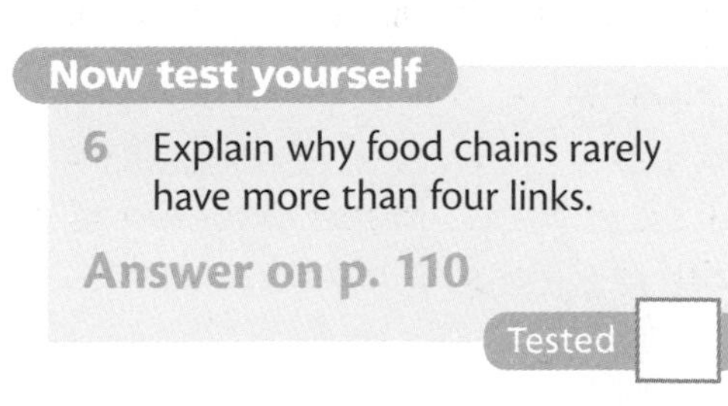

Now test yourself

6 Explain why food chains rarely have more than four links.

Answer on p. 110

Tested

Energy and food production

Net productivity

Revised

Plants capture sunlight energy and incorporate it into the chemical bonds of the carbohydrates they synthesise. Some of this energy is used by the plant for its own uses: remember that plants respire too. The energy available to the rest of the food chain is the energy locked into the starch, cellulose, sucrose and other substances when the plant is eaten. An important equation is:

net productivity = gross productivity – respiratory loss

where net productivity is the energy that is left in a plant after respiratory loss and gross productivity is the total energy captured by the plant in photosynthesis.

Plants require energy for their own metabolic processes, just like animals. Net productivity is important because it is 'profit' — the energy that goes to make new plant tissue. This is the energy available to an animal when it eats the plant and to a farmer it is the energy available in the crop.

Modern farming practices

Revised

Modern intensive farming practices have been developed to increase the efficiency of **energy conversion**. There are several methods, but the benefits of increased yields must be balanced against the financial costs of the process, the environmental costs and the welfare of the animals.

Fertilisers

Soil contains **humus**, a sticky mixture of rotting organic matter consisting of dead plants, animals and faeces. Good soil has a high humus content. The action of soil microbes releases mineral ions such as nitrate and phosphate, which is absorbed by plant roots. When a crop is grown in the same soil year on year, the ion content is depleted and it becomes a limiting factor in crop growth. Two types of **fertiliser** replace these ions:

- **natural (organic) fertilisers** — usually animal dung
- **artificial (inorganic) fertilisers** — a commercial product that usually comes in sacks

Table 3.1 Advantages and disadvantages of natural and artificial fertilisers

Type of fertiliser	Advantages	Disadvantages
Natural (organic)	Cheap Improves soil structure Improves soil moisture Releases nutrients slowly as it decomposes	Bulky Smelly Difficult to store and distribute May contain pathogens Ion content is difficult to control
Artificial (inorganic)	Ion content can be carefully controlled and matched to the needs of the soil/crop Easy to store and distribute	Expensive May cause osmotic damage May leach into waterways, causing eutrophication

Pesticides

Pesticides are **chemicals** that control **pests**. The ideal pesticide is:

- selective — it kills only the pest and nothing else
- biodegradable — it breaks down quickly in the environment, leaving harmless residues
- cheap, safe to handle and safe to store

Bioaccumulation can be a problem. Many pesticides are lipid-soluble, which means that they accumulate in the fatty tissues of plants and animals. Animals can excrete only water-soluble compounds, so any pesticide they take in builds up over time. At low levels, many pesticides are harmless to vertebrates, but the higher up the food chain you go, the more the pesticide builds up in the body of the individual. Eventually, a threshold is reached where the animal is either killed or made less fertile.

Bioaccumulation is the way in which some substances are concentrated in the bodies of animals — the higher the position of the animal in the food chain, the greater the concentration of the substance.

DDT is a classic example of a pesticide that accumulates higher up the food chain (Figure 3.5). When a large fish eats, say, 100 small fish, the DDT from all these accumulates in the body of the fish-eating bird. Each time, a little more DDT enters the body and accumulates in the tissues. Although eating a single contaminated small fish would have little effect, eating many of them leads to a build-up of concentration.

Examiner's tip

Strictly speaking, *bioaccumulation* refers to the build-up of pesticide within one organism, whereas *bioamplification* refers to the build-up from one trophic level to the next. However, in the exam it is the idea that is important.

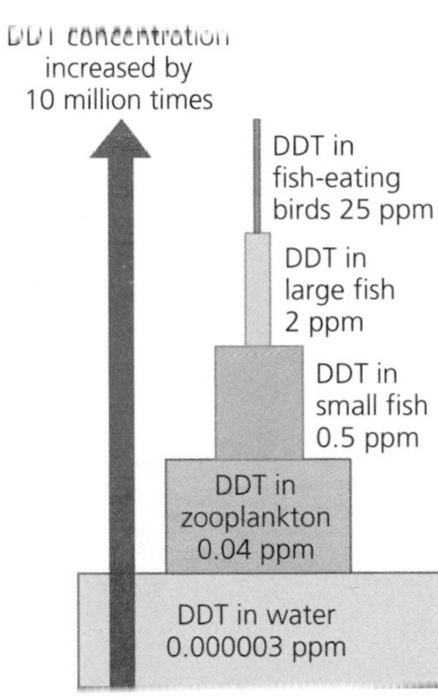

Figure 3.5 DDT is stored in fatty tissue, so each organism in the food chain receives the dose stored in all the animals it eats

Biological control

Biological control involves the use of a natural predator or disease. Examples include:

- using ladybirds in glasshouses to keep the numbers of aphids down
- using the *myxoma* virus to control the rabbit population

Biological control is the use of natural predators or parasites to control pest populations.

A perfect biological control agent needs applying just once, after which it keeps the pest population below the level at which it causes serious economic damage. A good biological control agent:

- attacks only the pest species
- does not carry diseases that might be introduced into the ecosystem
- can survive in the new environment, so it does not need constant reintroductions
- will not itself become a pest

An example of a less-than-perfect biological control agent is the introduction of cane toads in Australia, which did not eat the cane grubs that they were supposed to control. Instead, they ate virtually everything else they could, they had no predators, they out-competed native amphibians and they spread uncontrolled across the whole country.

Integrated pest-management systems

Integrated pest-management systems involve managing pests by making use of all available methods of control. Pests are managed in the most economical way and with the lowest possible risk to people and the environment. The life cycle of the pest is studied and then kept at a tolerable level. Cultural methods such as crop rotation can keep pests at a low level, with biological and/or chemical control being used as needed.

Intensive rearing of domestic livestock

All common farm animals — cattle, pigs, sheep and chickens — are warm blooded, so they expend a lot of energy in just maintaining body temperature. There are three common **intensive-rearing practices** that minimise energy loss so that more energy is used to make meat:

- minimise space — if animals cannot move around much, they save energy
- keep in heated barns — less energy is expended on maintaining core body temperature
- feed concentrates that contain less indigestible matter than normal feed, so less energy is lost in faeces — cows fed on corn put on weight faster than those fed on grass

Selective breeding has been carried out by farmers for centuries and it involves breeding from animals with desired characteristics, such as high milk yields in cattle and fine wool in merino sheep.

Examiner's tip

The only way that energy is lost from an organism is as heat, which is released during respiration. Reducing the rate of respiration saves energy.

Typical mistake

Many candidates say that 'energy is used for respiration', but it is not. Respiration *releases* energy.

Comparing natural ecosystems and modern farming

Revised

Intensive rearing of domestic livestock obviously uses more energy than traditional farming methods, but the benefits of increased productivity often outweigh the costs.

Nutrient cycles

The role of microorganisms

Revised

The atoms and molecules that make up our bodies have all been part of many other organisms before and will be part of many more in the future. Remember that elements are recycled, but energy is not. Energy from the sun drives the cycles. If the sun stopped shining, photosynthesis would stop and so would all life on Earth.

Nutrient cycles rely on the action of **saprobionts** (bacteria and fungi) in making the elements locked up in complex organic molecules available to plants once again. These saprobionts feed by **extracellular digestion**, which breaks down dead organic matter. They synthesise and secrete digestive enzymes onto the surface of their food. These enzymes break down complex molecules such as starch and protein into smaller, soluble ones that are then absorbed.

The carbon cycle

Revised

Carbon is a key component of all major biological molecules: carbohydrates, lipids, proteins, nucleic acids and many others. It is released into the atmosphere as **carbon dioxide** and dissolved in water as hydrogencarbonate (HCO_3^-) ions. This carbon is fixed into organic molecules by **photosynthesis**.

In the **carbon cycle** (Figure 3.6), carbon passes up the food chain, beginning when the plants are eaten. At every stage, some is released back into the atmosphere by **respiration**, which also loses energy as heat. Eventually, all of the carbon that is not released by respiration ends up in dead organic matter: dead leaves, animals and faeces. This carbon is released back into the atmosphere by the respiration of **decomposers**.

Sometimes, organic material does not rot because conditions are not right, e.g. there is not enough oxygen or the environment is too acidic. In this case, the material becomes fossilised and forms carbon-rich deposits such as coal, gas and oil. This carbon is released back into the atmosphere by **combustion**.

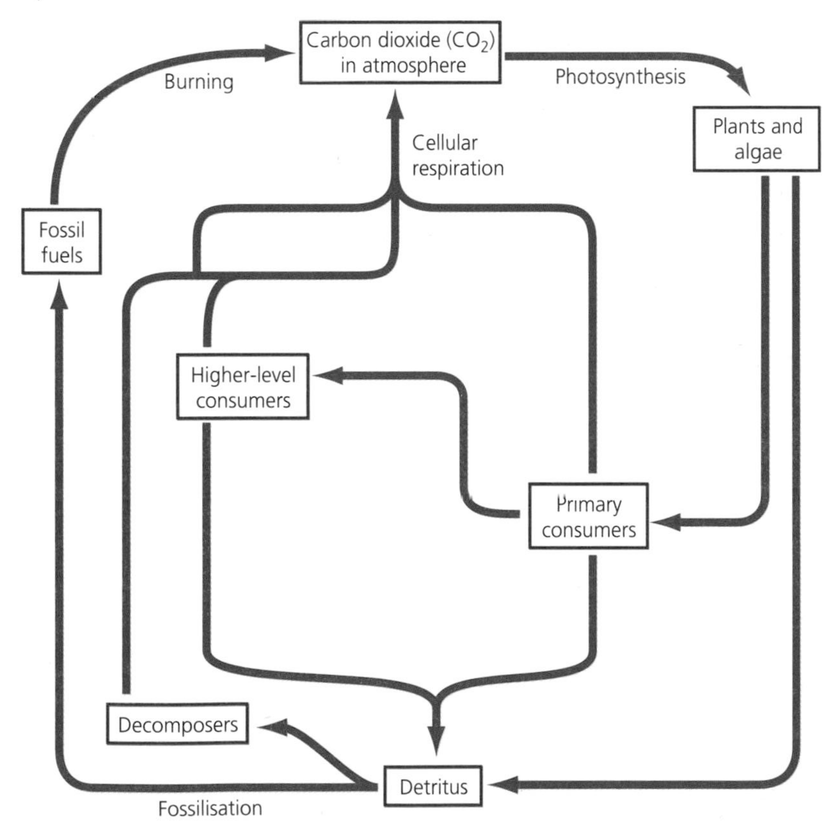

Figure 3.6 The carbon cycle is a balancing act between the carbon in carbon dioxide in the air and water and the carbon that is locked up inside organic molecules

Rotters, decomposers, saprophytes are all words for the same organisms. Saprobionts is the best word to use.

Typical mistake

Many candidates state that carbon enters plants via the roots, but it does not. It is absorbed as carbon dioxide through the stomata in the leaves.

Now test yourself

Tested

7 Name the only process that removes carbon dioxide from the atmosphere.

8 List two processes that return carbon dioxide to the atmosphere.

Answers on p. 110

The nitrogen cycle

Revised

Nitrogen is an essential component of proteins and nucleic acids (DNA and RNA), along with smaller molecules such as ATP, urea and ammonia. Some 80% of air is nitrogen gas (N_2), but the molecule is stable and not usually available to organisms. It takes a lot of energy to split molecules of nitrogen gas because the two atoms are connected by a strong triple bond.

Now test yourself

Tested

9 Explain why atmospheric nitrogen is not normally available to organisms.

Answer on p. 110

Plants absorb nitrogen as nitrate ions (NO_3^-). These ions are present in the water in soil and are absorbed into the roots by **active transport**. Plants combine nitrate with the substances made in photosynthesis to make amino acids, nucleotides that, in turn, make proteins and nucleic acids.

In the nitrogen cycle (Figure 3.7), nitrogen passes up the food chain in organic molecules. Animals get their nitrogen in proteins and nucleic acids. Eventually, all the nitrogen ends up in dead organic matter: dead plants and animals, faeces and urine. At this stage, the nitrogen is still locked up in organic molecules, so the action of saprobionts (bacteria and fungi) takes over. **Saprobiontic nutrition** involves the break-down of large molecules by extracellular digestion. Proteins are broken down into amino acids that the saprobionts absorb. These amino acids are **deaminated** and the resulting ammonium (NH_4^+) ions are released as a by-product in a process called **ammonification**.

The next stage is **nitrification**. There are two types of **nitrifying bacteria**, each of which play a part in converting ammonium into nitrate in a two-stage process: first, ammonium is oxidised into nitrite ions, then this nitrite is further oxidised into nitrate ions. Plants can then absorb the nitrate and the cycle continues.

However, there are two complications to the nitrogen cycle:

- **nitrogen fixation** — nitrogen-fixing bacteria contain the enzyme **nitrogenase**, which convert nitrogen gas into ammonium ions so that nitrogen can re-enter the cycle. Nitrogen gas can also be fixed by lightning during electrical storms
- **denitrification** — denitrifying bacteria lose nitrate from the cycle by converting nitrate ions into nitrogen gas. This process tends to occur in waterlogged, anaerobic soil

Deamination is the removal of the amino (NH_2) group from an amino acid.

Ammonification is the stage in the nitrogen cycle in which saprobiotic microorganisms break down organic, nitrogen-containing substances such as proteins and produce ammonium compounds.

Nitrification is an important stage in the nitrogen cycle in which ammonium compounds are converted to nitrites and nitrates.

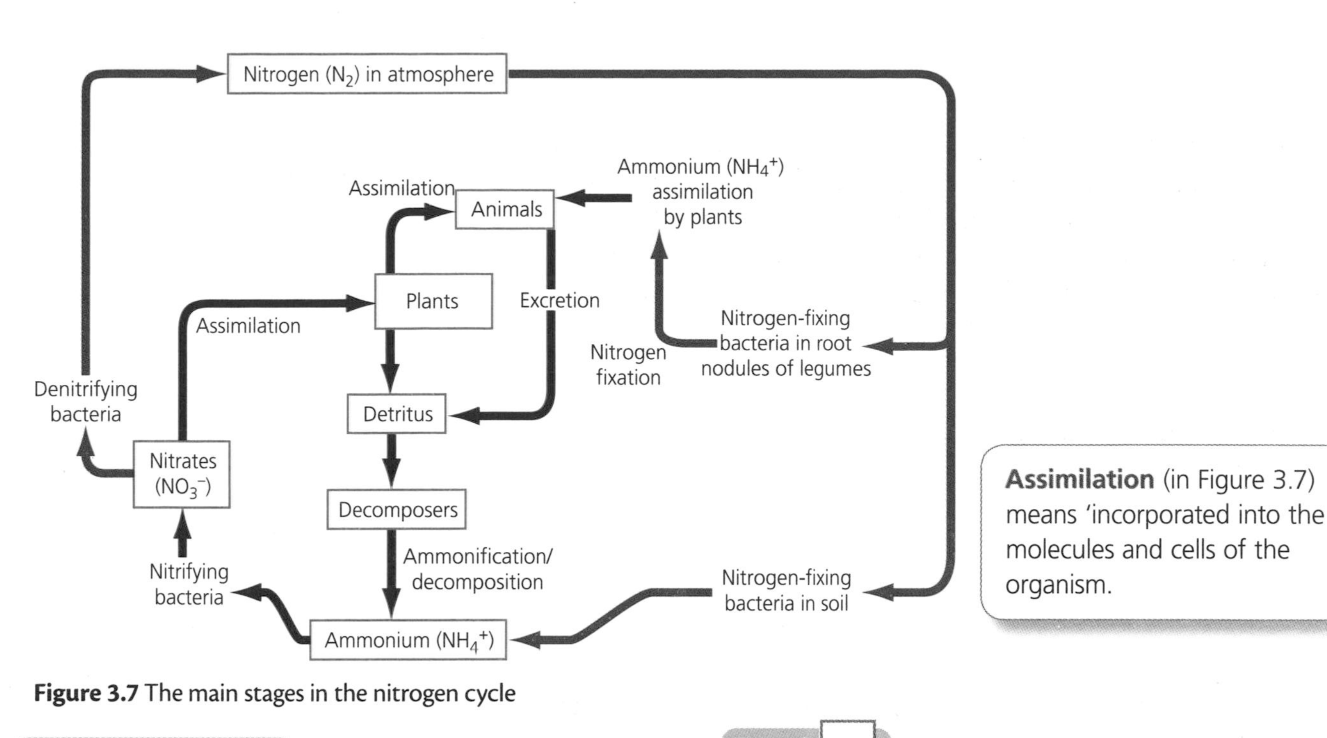

Figure 3.7 The main stages in the nitrogen cycle

Assimilation (in Figure 3.7) means 'incorporated into the molecules and cells of the organism.

Now test yourself Tested

10 Explain what is meant by the term *nitrogen fixation*.

11 List the two ways in which nitrogen can be fixed.

12 Some plant species, such as clover, beans and peas (legumes), have root nodules that contain nitrogen-fixing bacteria. Suggest the advantage of this to:

(a) the plant

(b) the bacteria

13 Nitrogen-fixing bacteria have an enzyme called nitrogenase, which has the ability to fix nitrogen gas into ammonia. Suggest why scientists want to put this enzyme into agricultural crops such as wheat and corn.

Answers on p. 110

Examiner's tip

The specification is very clear — you do *not* need to know the names of any of the bacteria in the nitrogen cycle. If your notes are covered with names like *Nitrosomonas* and *Nitrobacter*, get rid of them.

Carbon

Global carbon dioxide concentration

Revised

In order for the composition of the atmosphere to stay in equilibrium, the processes that return carbon to the atmosphere (respiration and combustion) must be balanced by the only process that removes it (photosynthesis).

Compared to oxygen and nitrogen, there is little carbon dioxide in the atmosphere: about 0.039% — just 390 molecules in every million (usually written as 390 ppm). The concentration of carbon dioxide was fairly stable until about 1850, when it began to rise dramatically (Figure 3.8). In 1960 it was below 320 ppm and has been rising steadily ever since (Figure 3.9). The main reason for this increase is combustion and specifically the burning of fossil fuels.

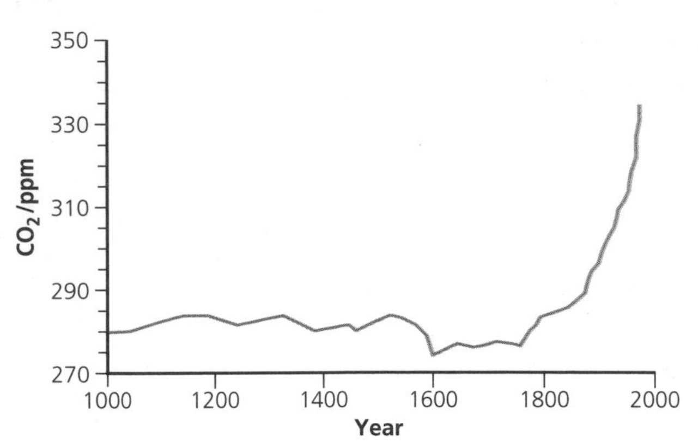

Figure 3.8 Changes in carbon dioxide concentration over the last 1000 years

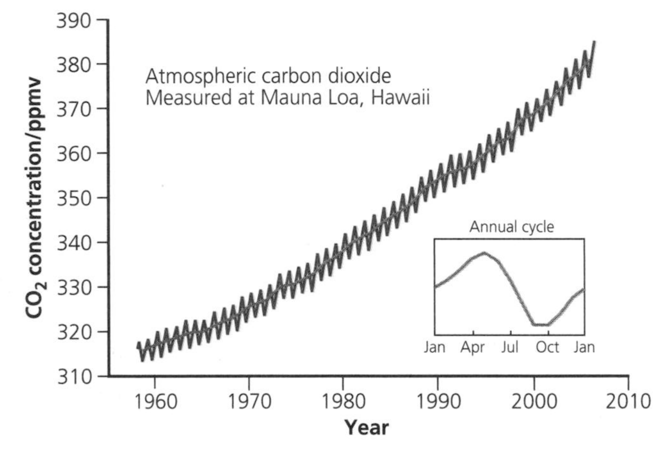

Figure 3.9 Changes in carbon dioxide concentration since 1960 at Mauna Loa

Making the problem worse is **deforestation**. There is a lot of carbon in wood, so when we cut down trees and burn them there is a lot more carbon dioxide in circulation.

Now test yourself Tested

14 Study Figure 3.9. Calculate the percentage increase in carbon dioxide concentration from 1962 to 2010.

Answer on p. 110

Global warming

Overall, the temperatures on Earth result from the balance between energy received and energy lost. The energy we get from the sun is not changing, but we are losing less because we are changing the composition of our atmosphere, allowing less radiation to escape. **Carbon dioxide** and **methane** are two major greenhouse gases, although **water vapour** and several other gases also contribute:

- Carbon dioxide results from respiration of living things and combustion.
- Methane comes from several sources, including the decomposition of waste, anaerobic respiration of wetlands and the metabolism of ruminants.

Now test yourself Tested

15 Explain how human activity causes global warming.

Answer on p. 110

The effects of global warming are many, complex and difficult to predict. You do not have to learn any specific examples, but the specification says that you should be able to interpret data about:

- the yield of crop plants
- the life cycles and numbers of insect pests
- the distribution and numbers of wild animals and plants

The study of the timing of natural events is called **phenology**. The life cycles of animals and plants have evolved together over a long period of time and are interdependent. Plants need to flower when the right species of insect is around to pollinate them. Hibernating animals need to wake up when there is enough food available and not before.

Now test yourself Tested

16 A warmer climate might cause a particular species of aphid to hatch earlier than its natural predator, the ladybird. Predict the consequences.

Answer on p. 110

Exam technique: interpreting data

Sample question

1 Study the following graph, which shows the rates of photosynthesis and respiration in a crop that is normally grown in a glasshouse.

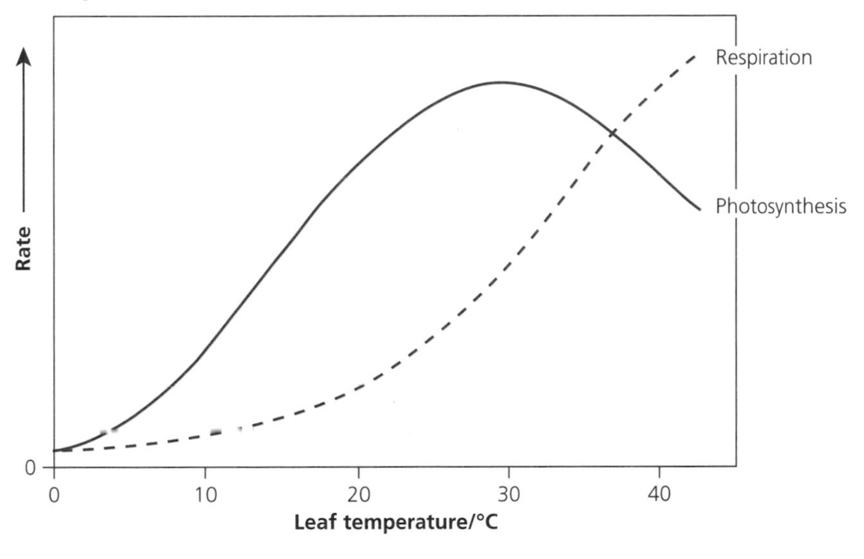

(a) Explain why respiration and photosynthesis are temperature-sensitive.

They are enzyme-controlled processes. (To be specific, the light-independent reaction of photosynthesis and the whole of respiration are enzyme-controlled and therefore temperature-sensitive. The light-dependent stage of photosynthesis, involving the excitation of chlorophyll, is not directly enzyme-dependent.)

(b) Explain how the rate of photosynthesis affects crop yield.

The higher the rate, the greater the yield (it makes organic molecules).

(c) Explain how the rate of respiration affects crop yield.

The higher the rate, the lower the yield (it uses organic molecules).

(d) Predict the temperature at which crop yield is greatest. Explain your answer.

About 25°C — this is the temperature at which photosynthesis exceeds respiration by the most.

(e) Predict the effect of an increase in temperature from the optimum to 3°C higher.

Loss in yield because of higher respiration.

Nitrogen

Leaching and eutrophication

Revised

Leaching is the loss of soluble substances such as nitrates and phosphates from the soil when water drains through it. When farmers apply **fertilisers** to their land, ions often get washed off by rainfall and find their way into rivers or lakes, where they may cause **eutrophication**. The nitrate, phosphate and other ions over-fertilise the aquatic ecosystem:

- The ions cause a bloom of algae because they reproduce faster than aquatic plants.
- The algae block the light, causing the death of the plants below.
- When the algae have used up all the nutrients they die, adding to the dead matter.
- Saprobiotic decay by aerobic bacteria uses up all the oxygen.
- A 'dead zone' results and all the native organisms that require oxygen die.

Eutrophication is an increase in the quantity of plant nutrients. The term is used when aquatic ecosystems are enriched with nitrates and phosphates, either as a result of the leaching of fertiliser from agricultural land or from sewage effluent.

Revision activity

Draw a flow diagram to summarise the stages involved in eutrophication.

Now test yourself

Tested

17 What evidence is there in the text that mineral ions are limiting factors in algae growth?

Answer on p. 110

Exam practice

1 Explain how the carbon in a dead leaf is made available for plant growth the following year. [4]

2 The following diagram shows the nitrogen cycle.

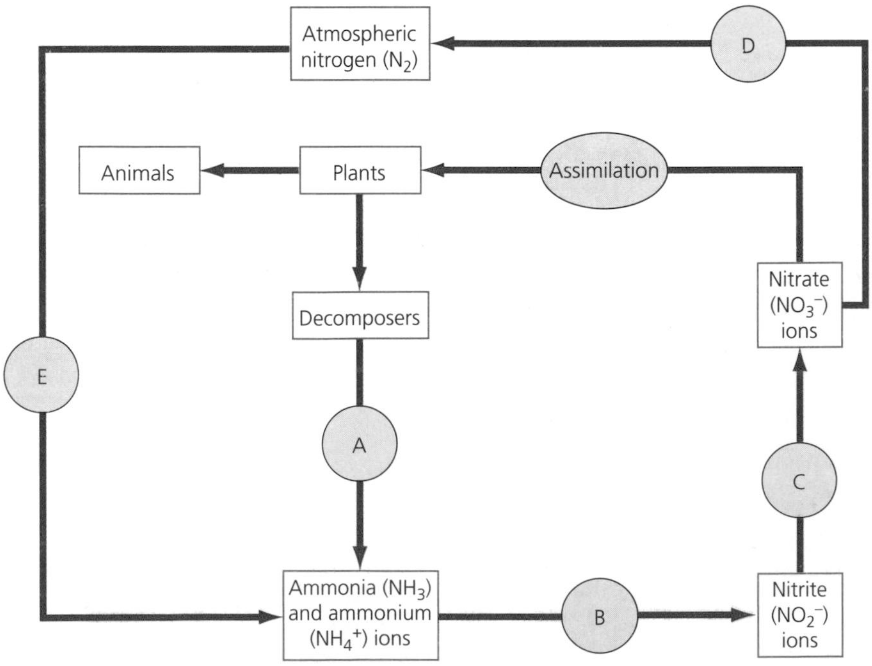

Atmospheric nitrogen (N_2)
D
Animals
Plants
Assimilation
Decomposers
Nitrate (NO_3^-) ions
E
A
C
Ammonia (NH_3) and ammonium (NH_4^+) ions
B
Nitrite (NO_2^-) ions

(a) Name a compound that is classed as organic nitrogen. [1]

(b) Name processes A, B, C, D and E. [2]

(c) Suggest why many crop rotation schemes include growing a legume such as beans. [3]

3 The following table shows the average dates of the first spawning of frogs on the south coast of England.

Year	Average date of first spawning
1955	16 February
1960	17 February
1965	13 February
1970	9 February
1975	7 February
1980	3 February
1985	1 February
1990	27 January
1995	23 January
2000	18 January

(a) Describe the pattern shown by the data in the table. [1]

(b) Suggest how the data would be different for ponds in Scotland. Explain your answer. [2]

(c) Suggest how the changing pattern may be a problem for the frog population. [2]

(d) Animals and plants that use temperature to synchronise their life cycles are affected by global warming. Suggest a more reliable environmental factor. [1]

Answers and quick quizzes online

Online

Examiner's summary

By the end of this chapter you should be able to understand:

- ✔ Energy enters an ecosystem when it is captured in photosynthesis, although a large percentage of sunlight is lost.
- ✔ All energy transfers are inefficient and energy is lost at each trophic level.
- ✔ Pyramids of numbers, biomass and energy are different quantitative ways of describing relationships between the different trophic levels in an ecosystem.
- ✔ Natural ecosystems can be compared with modern intensive farming in terms of energy input and productivity.
- ✔ Farming practices increase the efficiency of energy conversion. Pests of crops can be controlled by chemical or biological methods, and by integrated pest management (IPM). Energy losses from animals can be minimised by intensive rearing, though there are ethical issues.
- ✔ Microorganisms have an important role in the carbon and nitrogen cycles, which involve the basic processes of saprobiotic nutrition, ammonification, nitrification, nitrogen fixation and denitrification.
- ✔ Respiration, photosynthesis and human activity are important in changing global carbon dioxide concentration.
- ✔ Carbon dioxide and methane play a role in bringing about global warming.
- ✔ There are serious environmental issues arising from the use of fertilisers, including leaching and eutrophication.

4 How ecosystems develop

Succession

From pioneer species to climax community

Revised

How do we go from bare rock, sand or sterile soil to a fully developed, mature forest? The process involves **colonisation** and **succession**, as shown in Figure 4.1.

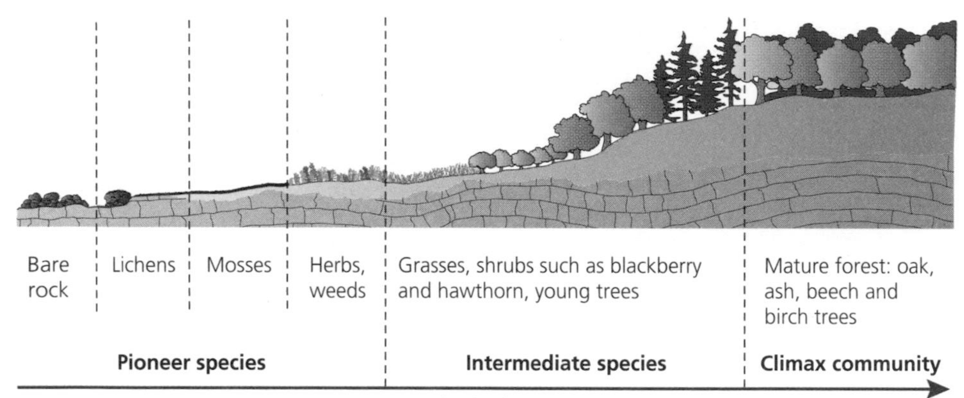

Figure 4.1 Succession from bare rock

In the beginning, there is a harsh **abiotic environment**. On land, nutrients and water are in short supply. Only a few specialist organisms can colonise: the **pioneer species**. When the pioneer species have become established, they make the environment more favourable and **less hostile**. This allows other organisms to survive, which then succeed the colonisers by **out-competing** them. Succession continues as the conditions become increasingly favourable, with different species dominating at each stage. Succession stops when a **climax community** is established.

A **climax community** is one in which the plants and animals remain stable and exist is balance with each other and their environment.

For example, starting with bare rock, the pioneer species are usually lichens — fungi that contain algae. This is a **mutualism** — an association in which both benefit. The fungi provide carbon dioxide and nitrogen as well as protecting the algae from drying out. In turn, the algae photosynthesise and provide organic molecules such as sugars. Each provides what the other needs, so they can survive in harsh places.

Once the lichens are established, they change the conditions so that simple plants such as mosses can get a hold. These mosses hold moisture and trap soil particles so that more complex plants such as ferns can become established. More and more soil accumulates under and between the plants, so that moisture and humus build up. Succession continues until the plants with woody stems dominate — the shrubs and trees. These large, tall plants out-compete the others for light and can survive from one year to the next, growing larger and larger.

In the UK, the climax community is **deciduous forest** — a stable population of dominant tree species that shed their leaves each year. Much of the UK was covered in this type of forest until humans cut it down. Globally, the climax community that develops depends on the climate. Examples include tropical rainforest, temperate coniferous forest and tundra.

Grazing

Revised

Succession cannot take place if there is a lot of grazing. Sheep, rabbits, cows and deer can all prevent the ecosystem from developing beyond grassland because they remove the growing tips of many plant species, including tree saplings. Grass has its growth point at the bottom of the stem, so if the tip is removed it still keeps growing.

Now test yourself

Tested

1. What is colonisation?
2. Why does succession happen?
3. What is a climax community?
4. What is humus?

Answers on p. 110

Management of succession

Revised

There are situations where the management of succession brings benefits, such as:

- maximising the growth of trees
- maximising habitat **diversity**
- maintaining the look of the countryside

For example:

- the regular burning of grouse moors encourages new growth of heather shoots that is a perfect habitat for grouse
- coppicing (cutting down some large trees to keep the succession at an intermediate stage) can stimulate new tree growth, creating more niches for endangered species
- in the Lake District, sheep farming is encouraged because other practices would dramatically change the landscape — the look of the countryside is important for tourism

Exam practice

1 Define the term *biodiversity*. [2]

2 Biodiversity generally increases as an ecosystem develops. Explain why. [3]

3 Coppicing is a practice that cuts down trees in a systematic manner so that they cannot grow to their full height, but which encourages growth of new shoots and smaller shrubs. Explain the conservation benefits of coppicing. [3]

Answers and quick quizzes online

Online

Examiner's summary

By the end of this chapter you should be able to understand:

- ✔ Ecosystems develop by a process of colonisation by pioneer species followed by succession that ends with a climax community.
- ✔ At each stage in succession, certain key species change the environment so that it becomes more suitable for other species.
- ✔ The changes in the abiotic environment result in a less hostile environment and changing diversity.
- ✔ Conservation of habitats frequently involves management of succession.

5 Genetics and speciation

Inheritance

Key concepts

Revised

A **genotype** is the **genetic constitution** of an **organism**. A **phenotype** is the observable characteristics of an organism as a result of the **alleles** it has inherited and its environment.

Alleles are alternative versions of the same gene. For example, a particular species of plant might have a gene for petal colour that has two alleles: one codes for purple flowers and one for red flowers. New alleles are created by the process of **mutation**.

A specific gene always occurs at the same position on a chromosome. This is called its **locus** (plural: loci).

Organisms usually have two copies of each gene because chromosomes come in pairs. If the alleles at a specific locus are the same, they are **homozygous**, written as AA or aa, for example. If they are different, they are **heterozygous** (Aa).

Alleles are different forms of a particular gene. Some genes have two or more alleles. For example, there might be a gene for coat colour in rats, with a *B* allele for brown fur and a *b* allele for albino fur.

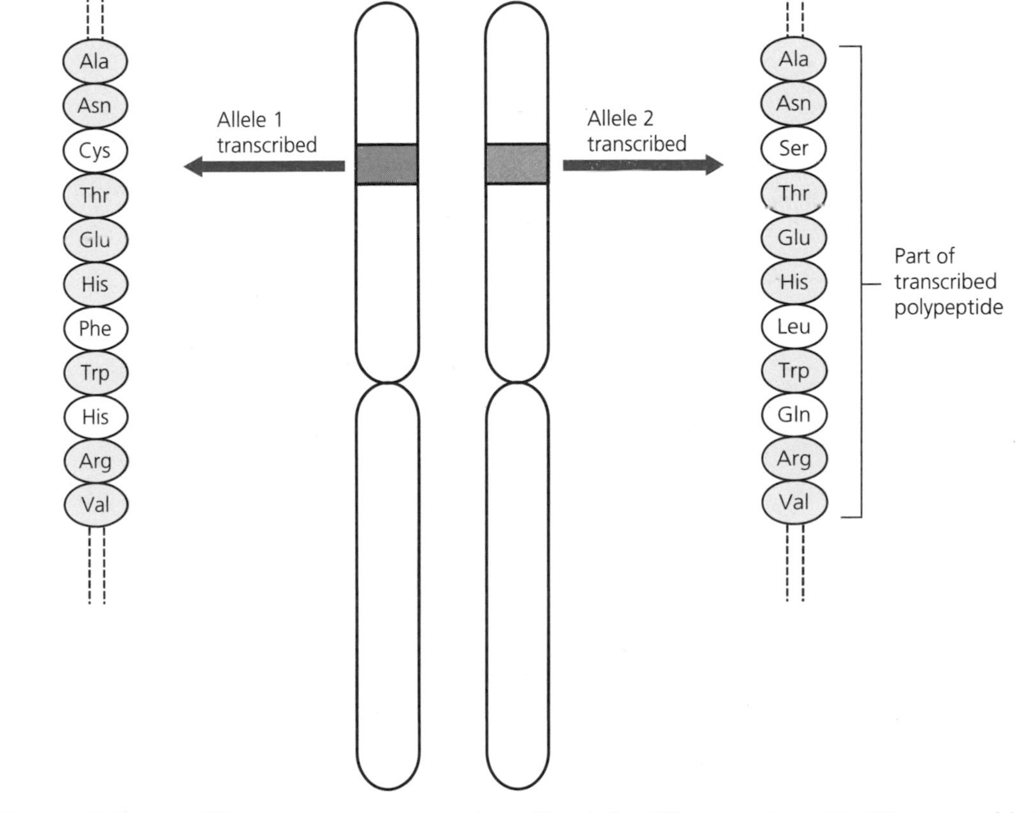

Figure 5.1 Different alleles have different base sequences and so will code for different amino acids. When assembled, the amino acid chain will fold and bend into a different polypeptide or protein

Dominant alleles are those which, if present, are expressed in the phenotype. **Recessive alleles** are only expressed when no dominant allele is present.

Now test yourself

1 Define the term *allele*.

Answer on p. 110

Tested

Monohybrid crosses

Revised

Monohybrid crosses involve **single genes**, usually with two alleles. In this example we will look at cystic fibrosis, an inherited genetic disorder caused by single faulty recessive allele. This means that it must be inherited from both parents (Figure 5.2). The F allele is dominant, so if a person has this allele it does not matter what the other one is — the f allele does not produce any effect.

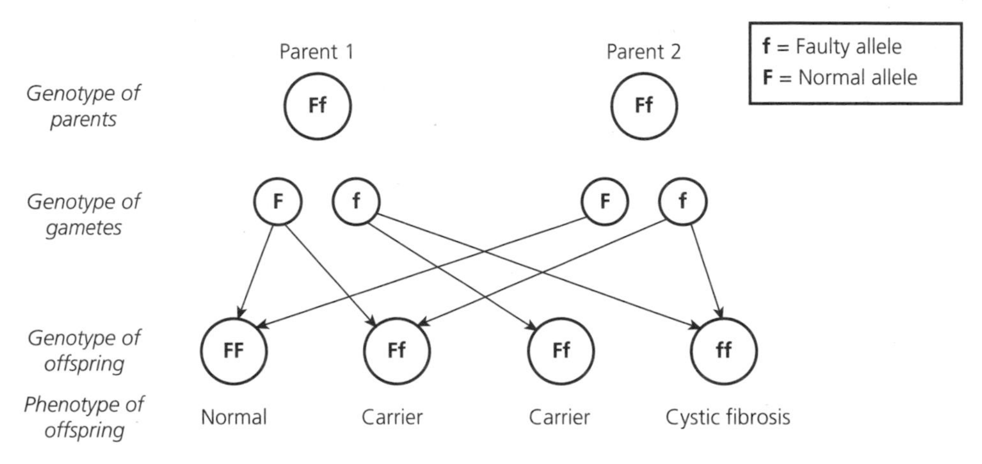

Figure 5.2 Genetic diagram to show the inheritance of cystic fibrosis

Notice that the offspring with Ff are labelled 'carriers'. A carrier has one copy of the faulty allele but does not have the disorder themselves. In this example, both parents are carriers, giving rise to a 1 in 4 or 25% chance that each child will be born with cystic fibrosis. Probability should be expressed as a decimal, in this case 0.25.

Examiner's tip

Candidates often lose marks by expressing the results in the wrong form. A 3:1 ratio, 1 in 4, 0.25 and 25% are all the same thing, so make sure you give the answer in the form required by the question.

Now test yourself

Tested

2 If the parents are FF and Ff, what is the probability that their first child will be born with cystic fibrosis?

Answer on p. 110

Codominance

Revised

Codominance is seen where there is no dominant and no recessive characteristic, so both alleles are expressed in the phenotype. For example, snapdragons have codominant alleles for red flowers (genotype RR) and white flowers (genotype WW). If both alleles are present, the flowers are pink (genotype RW). Both alleles R and W are equally dominant (Figure 5.3).

Codominance is seen where both alleles contribute to the phenotype.

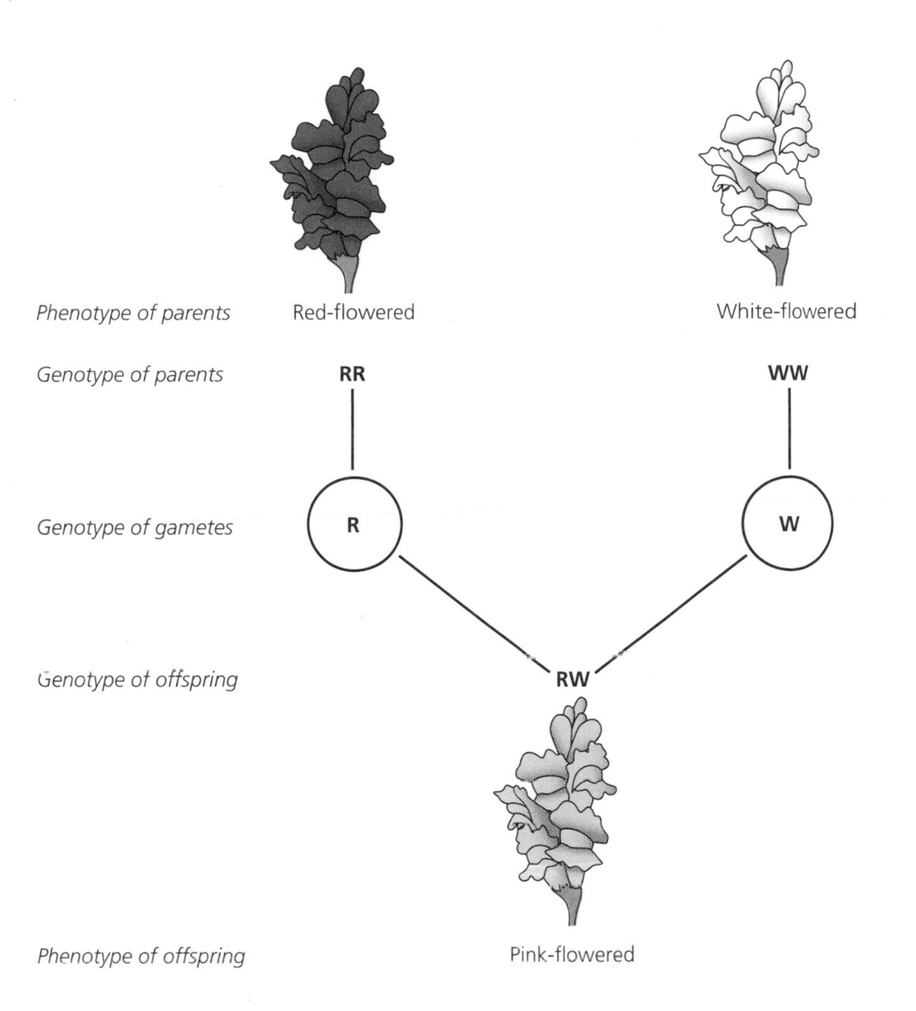

Figure 5.3 Codominance in snapdragons

Now test yourself

Tested

3 Define the term *codominance*.

4 If two pink snapdragons were bred together, predict the colour and ratios of the offspring.

Answers on p. 110

Multiple alleles

Revised

Having studied genetics, it is tempting to think that all genes have two alleles, but they do not:

- some genes have no alleles — there is just one form because the gene codes for a vital protein
- some genes have two alleles, as we have seen
- some genes have multiple (more than two) alleles

For example, the gene that codes for the human ABO blood group has three alleles, I^A, I^B and I^O. I^A and I^B are codominant over I^O. Each individual has two copies of these alleles, so:

- genotypes $I^A\,I^O$ or $I^A\,I^A$ will produce blood group A
- genotypes $I^B\,I^O$ or $I^B\,I^B$ will produce blood group B
- genotype $I^O\,I^O$ will produce blood group O
- because of the codominance, genotype of $I^A\,I^B$ will produce blood group AB

Now test yourself

5 A child has a mother with blood group AB and a father with group O. List the possible blood groups of the child.

Answer on p. 110

Tested

Family tree diagrams

Revised

Family tree or pedigree diagrams are popular in exam questions. Figure 5.4 shows the inheritance of albinism in a particular family. Albinos have the genotype aa — they cannot make normal skin pigmentation.

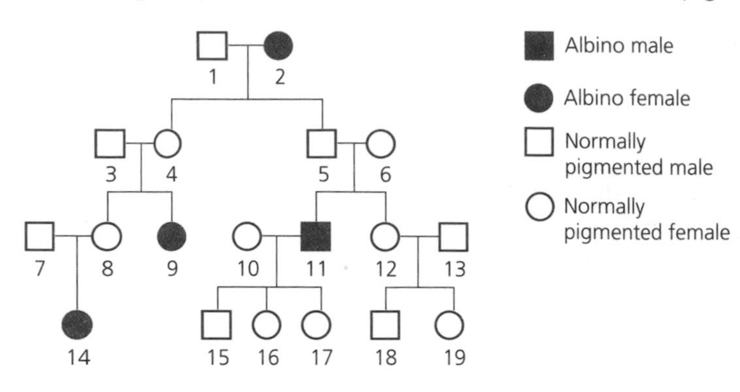

Figure 5.4 Albinism

If A is the allele for normal pigmentation and a is the albino allele, we can work out the genotypes of many family members. For example, we know that individual 2 must be aa, whereas individual 1 could be AA or Aa.

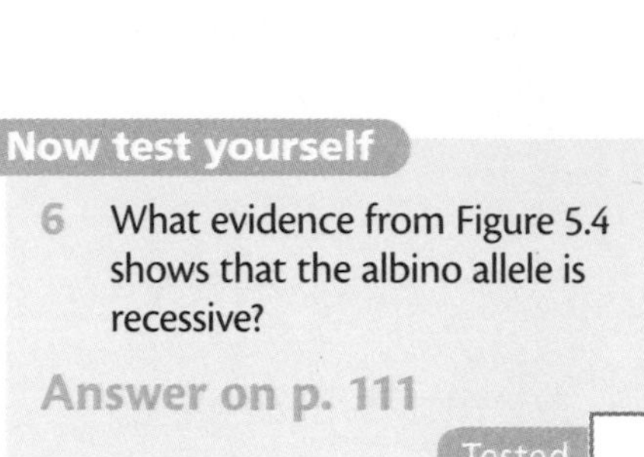

Now test yourself

6 What evidence from Figure 5.4 shows that the albino allele is recessive?

Answer on p. 111

Tested

Sex-linked characteristics

Revised

Figure 5.5 shows a complete set of human chromosomes. There are 23 pairs; 22 are called **autosomes** and they are **homologous** because they have the same genes at the same loci, although they may not have the same alleles.

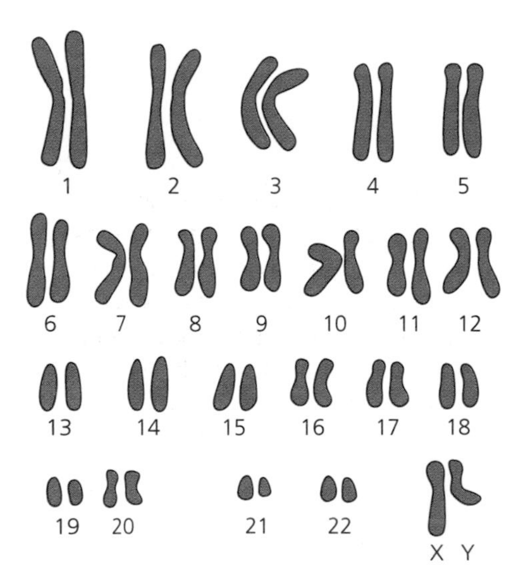

Figure 5.5 A karyotype — the full set of human chromosomes

The last pair of chromosomes are the sex chromosomes. Females have two large X chromosomes that contain thousands of genes and they are also homologous. If a gene is found on the sex chromosomes, it is said to be **sex-linked**. Of these, there are far more X-linked genes than Y-linked genes because the X chromosome is larger.

Males have only one copy of X-linked genes, so that allele is always expressed — there is no second (dominant) copy to mask its effects. Males cannot be carriers. If they are unlucky enough to have a faulty allele, they will have the disease that the allele causes. Examples of sex-linked disorders include haemophilia, colour blindness and Duschenne muscular dystrophy. All are X-linked, so males are more likely to have them.

Now test yourself

Tested

7 What is a sex-linked gene?

8 Explain why males are more likely than females to suffer from sex-linked disorders.

Answers on p. 111

With sex-linked genotypes, we show the chromosome and the allele. In this example we will look at haemophilia, an inherited genetic disorder that impairs the body's ability to control blood clotting and coagulation.

H = normal allele

h = faulty allele

So, females have three possible genotypes:

X^HX^H = normal, healthy

X^HX^h = healthy but a carrier

X^hX^h = haemophiliac

Males have two possible genotypes. There is no allele on the Y chromosome:

X^HY = normal, healthy

X^hY = haemophiliac

In order for two healthy parents to have a haemophiliac son, the mother must be a carrier, so we know what the parent's genotypes must be.

Genotype of parents:	X^HX^h	X^HY		
Genotype of gametes:	X^HX^h	X^HY		
Genotype of offspring:	X^HX^H	X^HX^h	X^HY	X^hY
Phenotype of offspring:	healthy	carrier	healthy	haemophiliac
	female	female	male	male

Examiner's tip

Candidates often add random X and Y chromosomes when there is no need. If the question is about sex-linkage, it will clearly say so. If not, it isn't.

Examiner's tip

Not all species have XX females and XY males. Chickens, for example, are the other way around. The basic genetics is just the same, but the sexes are reversed. In this case, it is the females that are much more likely to suffer from sex-linked characteristics.

Now test yourself

Tested

9 Is it possible to have a female haemophiliac? Explain your answer.

Answer on p. 111

Sex-linked pedigree diagrams

When trying to make sense of pedigree diagrams involving sex-linked traits, the golden rule is all males get their Y chromosome from their father — otherwise they would not be males — and so their X chromosome must have come from their mother. Figure 5.6 shows the inheritance of sex-linked colour blindness in one family. Allele B for normal vision is dominant to allele b for colour blindness.

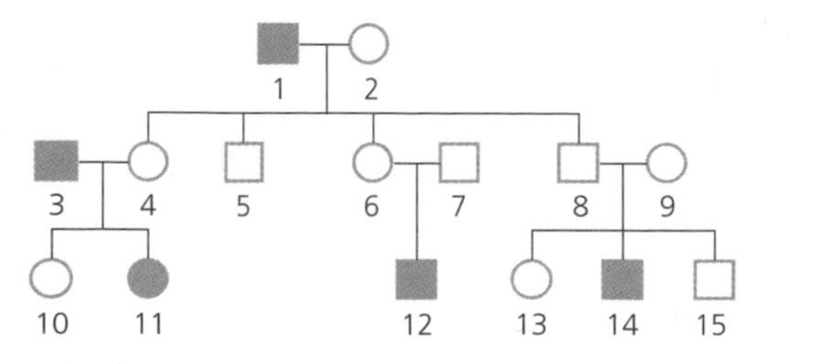

Figure 5.6 Inheritance of red-green colour blindness in one family

Now test yourself

Tested

10 Work out the genotypes of each individual in Figure 5.6 and give a brief reason for each. Use a table like this for your answer.

Individual	Genotype	Reason
1	X^bY	He is a colour-blind male
2		
3		

Answer on p. 111

The Hardy-Weinberg principle

Key concepts

Revised

The Hardy-Weinberg principle is a calculation that allows you to:

- calculate **allele frequencies** in a **population** starting from simple observations
- make predictions, e.g. is selection taking place, and is the population evolving?

Allele frequency is a measure of how common an allele is in a population. It is expressed as a decimal, so a frequency of 1 means 100%. If a gene has two alleles, A and a, their frequencies must add up to 1. If the frequency of A is 0.7 (70%), we know that the frequency of a must be 0.3 (30%).

A population contains individuals of three genotypes: AA, Aa and aa, where A is the dominant allele and a is the recessive allele. The Hardy-Weinberg equation is as follows:

$$p^2 + 2pq + q^2 = 1$$

where p^2 is the frequency of homozygous dominant individuals (AA), $2pq$ is the frequency of heterozygous individuals (Aa) and q^2 is the frequency of homozygous recessive individuals (aa).

A population with a large **gene pool** contains a lot of individuals with different genotypes. There are many alleles and a lot of **outbreeding** in which genetically different individuals reproduce together. If conditions become unfavourable, there is a good chance that some individuals will have favourable genotypes that will allow them to survive. Faulty alleles are rarely paired up, so few individuals suffer from genetic disorders.

A **gene pool** is all the alleles present in a particular population at a given time.

Revision activity

Rewrite the paragraph on the left, reversing each point to illustrate the other extreme.

Now test yourself

Tested

11 Define the term *gene pool*.

Answer on p. 111

Worked example of the Hardy-Weinberg principle

Revised

In this example, we will look at coat colour in mice. This is controlled by one gene and two alleles. Allele A codes for agouti fur (the normal colouration of wild mice), which is dominant over allele a, which codes for no pigment. So, AA and Aa mice are agouti, whereas aa mice are albino.

Starting with the simple observation that 16% of mice are albino, we can work out how many of them are AA and how many are Aa, despite the fact that the agouti mice all look the same. In a population where 16% of individuals show the recessive feature, this is a frequency of 0.16. Therefore:

$$q^2 = 0.16$$

so

$$q = \sqrt{0.16}$$
$$= 0.4$$

As $q = 0.4$, therefore p must be 0.6 because they must add up to 1. Knowing the values of p and q, we can calculate the frequencies of the different genotypes in the population:

The frequency of the homozygous genotype is:

$$p^2 = 0.6^2 = 0.36$$

The frequency of the heterozygous genotype is:

$$2pq = 2 \times 0.6 \times 0.4 = 0.48$$

In Hardy-Weinberg questions, many candidates get confused and end up scribbling formulae and numbers all over the place. It may help to draw a grid similar to the one in Table 5.2, either in rough or at the side of the question. This will help you to focus your thoughts on the information you have been given.

Table 5.2 Sample grid for Hardy-Weinberg calculations

Genotype	Phenotype	HW equation	Value
AA	Agouti	p^2	
Aa	Agouti	$2pq$	
aa	Albino	q^2	0.16

Examiner's tip

The frequency of the heterozygous individuals is $2pq$ because there are two ways of getting to that genotype. A from the male, a from the female or vice vesa.

Examiner's tip

Hardy-Weinberg questions often give you nice numbers to work with. If you find yourself with values like 0.16, 0.36 or 0.49, smile — it's easy to find the square root.

Examiner's tip

Some questions may give you allele frequencies. This makes life easier because you don't have to work out the square root of the q^2 value, but the simplicity of the calculations seems to confuse many candidates.

Now test yourself

Tested

12 The frequency of one allele is 0.78 and the frequency of a different one is 0.65. How do we know they cannot be alleles of the same gene?

Answer on p. 111

Predictions

Revised

The Hardy-Weinberg principle can also be used to see if a population or species is evolving or not. The principle predicts that allele frequencies will not change from generation to generation, unless:

- there is selection taking place — some genotypes may give individuals a greater chance of survival
- there is emigration or immigration, when individuals join or leave the population
- the population is small, so chance plays a large part
- there has been a mutation, creating one or more new alleles

A change in allele frequency from one generation to the next is the basis for **evolution**.

Selection

What is selection?

Revised

Selection occurs when an individual has a genotype that gives it an advantage, which means that they pass on more of their alleles to the next generation. Individuals that pass on a lot of their alleles are said to be 'fit'. The frequencies of these beneficial alleles will therefore increase in the next generation.

Typical mistake

Don't use the term 'survival of the fittest' — it is a popular term that makes no sense to a proper biologist. In biology, fitness is a measure of reproductive success, so a 'fit' organism is one that passes on more of its alleles than other organisms. 'Survival of the best reproducers' is a bit silly.

Examiner's tip

The examiners have chosen to use the expression *selection* in place of the more familiar *natural selection*.

Types of selection

Revised

There are two types of natural selection:

- **Directional selection** occurs when individuals with extremes of phenotype have an advantage, such as the fastest, largest or most tolerant to cold. As a result, one phenotype becomes rare and an alternative becomes common.
- **Stabilising section** occurs when individuals with extremes of phenotype are at a disadvantage. In this case, individuals with intermediate phenotypes are more likely to pass on their alleles to their offspring. An example of this is birth weight in mammals. Babies born too small can have all sorts of problems, including susceptibility to cold. Particularly large babies can cause problems during birth.

In Figure 5.7, the upper graph shows a variation in one factor. The mode (most frequent value) is marked in red. The graph represents the frequency distribution of this population before directional selection has occurred. The lower graph shows the same population after directional selection. As you can see, selection favours the individuals at one extreme, so in the following generations the mean shifts to the right.

In Figure 5.8, starting with the same graph as in Figure 5.7, stabilising selection selects against both extremes, resulting in the same mean value but less variation.

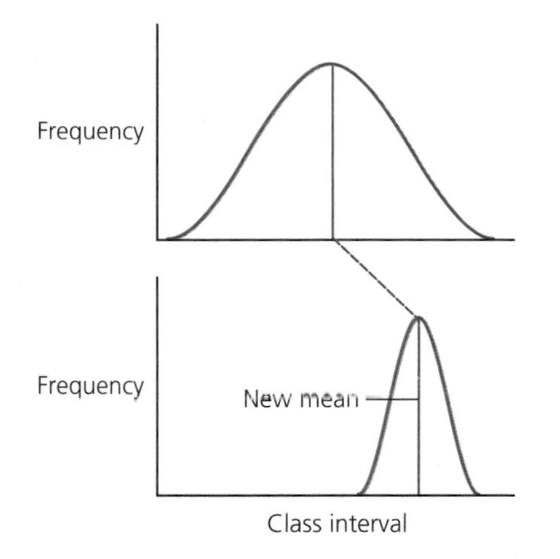

Figure 5.7 Directional selection

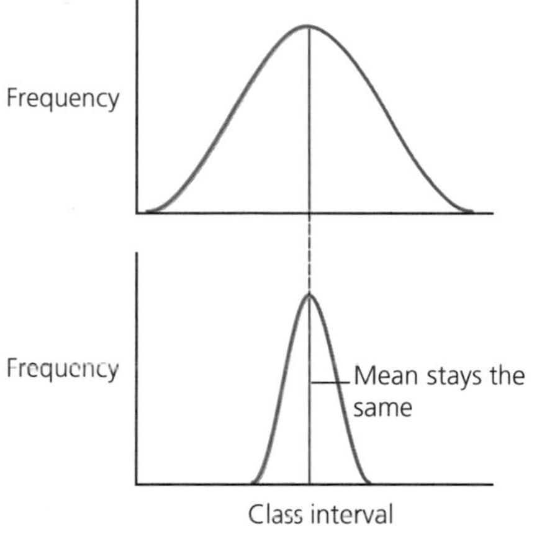

Figure 5.8 Stabilising selection

Speciation

How do new species develop?

Revised

Five steps are involved in the development of new species from a common ancestor:

1. The original population splits and the two or more populations become **isolated** from each other so that they cannot interbreed.
2. There is variation within the populations — a variety of different genotypes.
3. Selection acts differently on the two populations.
4. Allele frequencies change in the two gene pools.
5. Over the generations, genetic differences **accumulate** so that even if the two populations were to mix they could not interbreed.

Geographic separation of species

Revised

New species cannot develop unless two populations become isolated. There are lots of different types of isolating mechanisms, but the easiest one to understand, and the one you need to know, is **geographic isolation**, which occurs when two populations cannot physically contact each other. For example, if a road is built through a forest, this might separate two species of insect.

Examiner's tip

Don't go into too much detail about isolation. Geographic isolation is the only example you need.

Exam practice

1 In a particular species of beetle, the allele for green wing-cases is dominant over the allele for black wing-cases. In a population there were 400 individuals, of which 64 had black wing-cases.

(a) Calculate the percentage of beetles that are heterozygous. [4]

(b) The Hardy-Weinberg principle predicts that allele frequencies will not change from year to year, but scientists found that the frequency of the black wing-cased beetles was decreasing. Suggest an explanation. [2]

2 Explain how two species may evolve from a common ancestor. [4]

Answers and quick quizzes online

Online

Examiner's summary

By the end of this chapter you should be able to understand:

- ✔ An organism's phenotype is a result of the genotype and its interaction with the environment.
- ✔ Alleles are one or more alternative versions of the same gene. They may be dominant, recessive or codominant.
- ✔ Some genes are sex-linked, in which case the XX female will have two copies whereas males have only one. This affects the pattern of inheritance.
- ✔ The Hardy-Weinberg principle allows allele and genotype frequencies to be calculated using the formula $p^2 + 2pq + q^2 = 1$.
- ✔ The Hardy-Weinberg principle predicts that allele frequencies will not change from one generation to the next, unless there is selection, migration, mutation or a small population.
- ✔ Natural selection occurs when some genotypes have greater reproductive success than others. This affects allele frequency within a gene pool. Selection can be directional, producing change by selecting for an extreme of phenotype, or stabilising, which selects against the extremes.
- ✔ New species arise as a result of reproductive isolation followed by different selection pressures on the different populations. Over the generations, genetic differences accumulate so the populations cannot interbreed.

6 Stimulus and response

Survival and response

Responding to changes

Revised

All organisms have the same aim in life: to try to get food and to avoid being food for something else until they have had a chance to reproduce and pass on their alleles. Organisms increase their chance of survival by **detecting and responding to changes** in their environment.

A **stimulus** is a change in the environment that can be detected and a **receptor** is a specialised cell that can detect a stimulus. Large, complicated organisms such as humans have many different receptors for lots of different stimuli. For single-celled organisms, the whole organism is the receptor.

Now test yourself

1 Name five different stimuli that humans can detect.

Answer on p. 111

Tested

Tropisms

Revised

Plants have no nerves or muscles, so they cannot respond as fast as animals. However, they are sensitive to the environment and can respond by growing in a particular direction. Growth responses to **directional stimuli** in plants are called **tropisms** and these are achieved by changes in cell division and enlargement. Their purpose is to maintain the roots and shoots of plants in a favourable environment. There are many tropisms, such as:

- phototropism — the response to light. Stems are usually positively phototropic and grow towards the light, whereas roots are generally negatively phototropic and grow away from the light
- geotropism or gravitropism — both names for the response to gravity. Roots are usually positively geotropic and grow down into the soil, whereas stems are generally negatively geotropic and grow upwards
- hydrotropism — the response to water
- chemotropism — the response to chemicals
- thigmotropism — the response to touch

Tropisms are growth responses made by plants in response to an external stimulus.

Now test yourself

2 Explain the term *tropism*.

Answer on p. 111

Tested

Many aspects of plant life such as growth, flowering and leaf fall are controlled by **plant growth substances** known as **auxins**. The example you have to know is the control of phototropism by an auxin called **indoleacetic acid (IAA)**. Figure 6.1 explains the phototropic response.

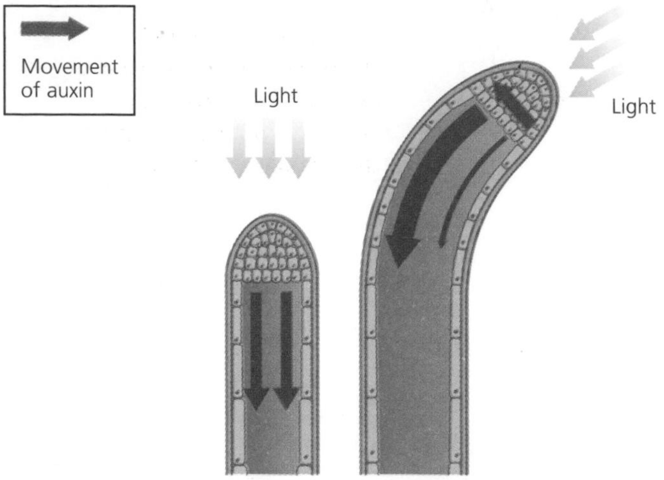

Movement of auxin

Light

Light

Figure 6.1 The mechanism of phototropism

1 Light illuminates the plant from one side.
2 The tip of the plant makes auxin, which is transported to the shaded side.
3 The auxin diffuses down the shaded side.
4 The auxin stimulates cell division and elongation.
5 The increased growth on the shaded side causes the stem to bend towards the light.

Taxes and kineses

Revised

At their most basic level, responses consist of detecting particular stimuli and then moving or growing towards or away from them. For organisms that can move, such as bacteria, algae and most animals, **taxes** (singular: taxis) and **kineses** (singular: kinesis) are common responses.

A taxis is a directional response that results from an organism being able to tell the direction of a stimulus. If an organism moves towards a stimulus, it is a positive taxis. For example, some bacteria are positively **aerotactic** because they move towards oxygen.

Some organisms can find favourable conditions even when they cannot detect in which direction to go using a kinesis, a non-directional response. Woodlice, for example, prefer dark, damp conditions. When in light, dry conditions they move quickly, making many turns. In this way they explore their environment. When, by chance, they find the conditions they prefer, they slow down and turn less frequently or simply stop.

The apparatus shown in Figure 6.2 is suitable for studying taxes and kineses in small invertebrates such as woodlice and maggots.

Now test yourself

3 Explain the terms *taxes* and *kineses*.
4 Some single-celled algae have flagella so they can move and are said to be positively phototactic.
   (a) Explain what the term *positively phototactic* means.
   (b) Explain the advantage of this characteristic to the algae.
5 Suggest the advantage to woodlice of seeking out damp, dark conditions.

Answers on p. 111

Tested

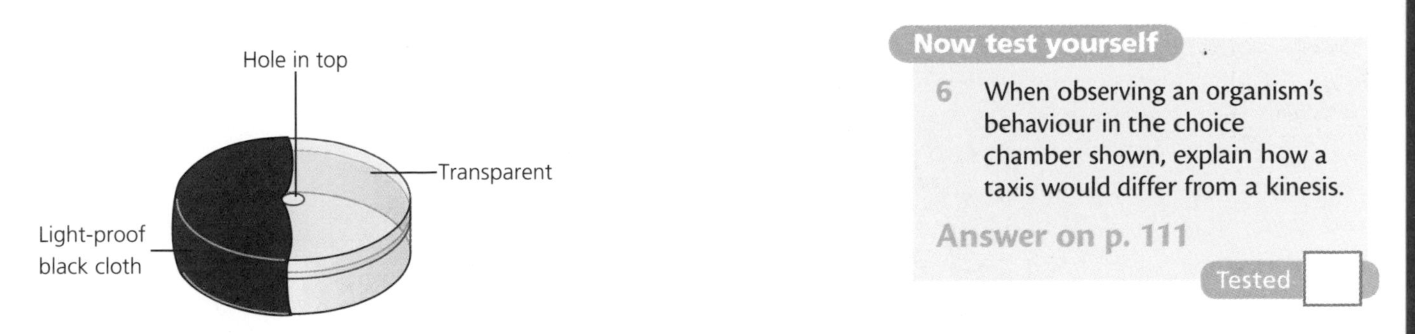

Figure 6.2 A simple choice chamber to investigate taxes or kineses in response to light

Now test yourself

6 When observing an organism's behaviour in the choice chamber shown, explain how a taxis would differ from a kinesis.

Answer on p. 111

Tested

Reflex arcs

Revised

Vertebrates have a **central nervous system (CNS)** consisting of a brain and a spinal cord. The basic sequence of events is shown in Figure 6.3.

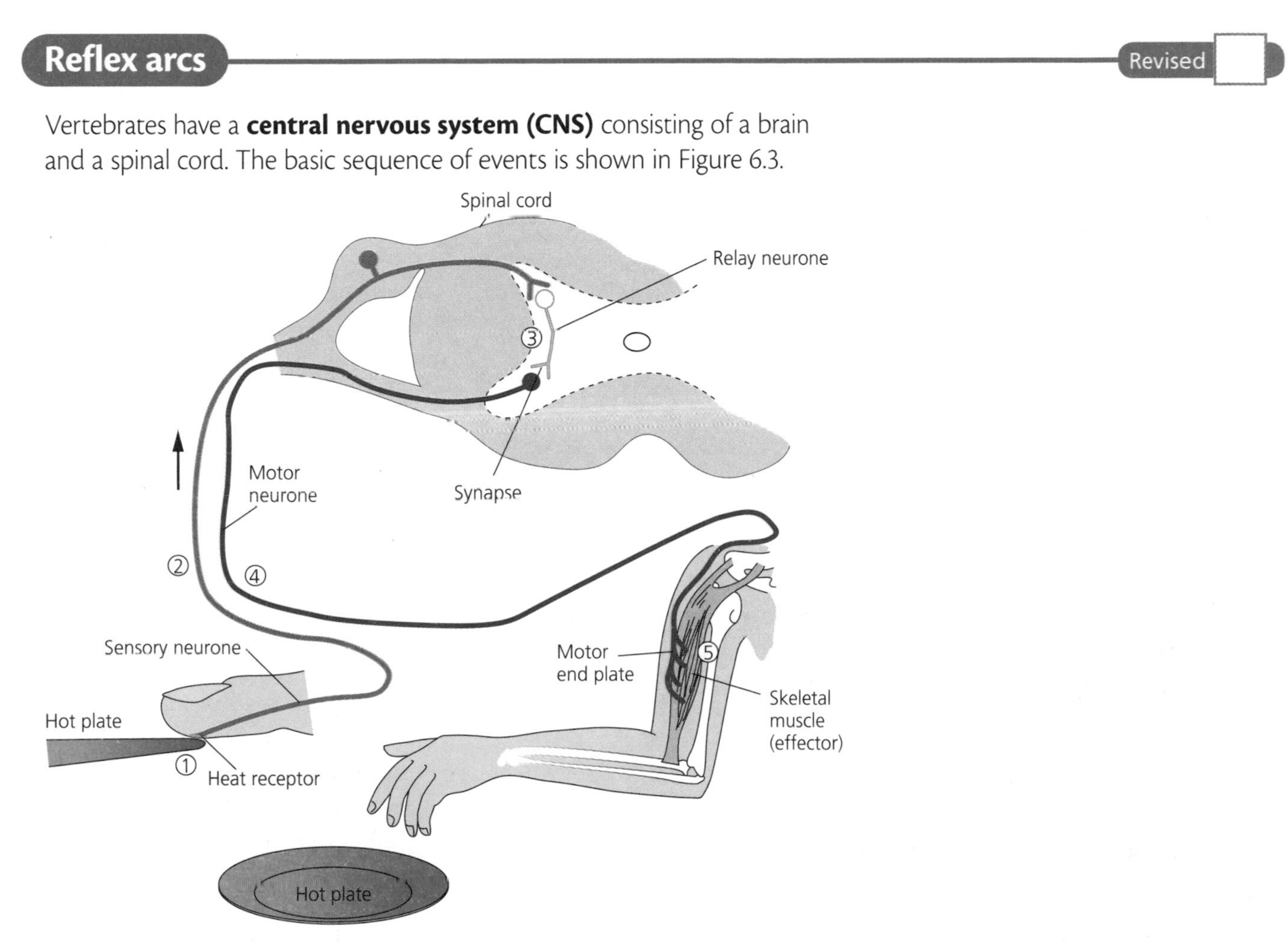

Figure 6.3 Withdrawing your hand from heat is an example of a spinal reflex, a reflex arc.

1. **Receptors** gather information by detecting stimuli from within the body and the external environment.
2. Sensory information passes into the CNS via impulses along sensory **neurones**.
3. The CNS processes this information and coordinates a response.
4. Impulses pass to **effectors** along motor neurones.
5. The effectors — muscles or glands — bring about the response.

Neurones are nerve cells that transmit nerve impulses.

The simplest response we can study is the **reflex arc**, which involves three neurones.

- a **sensory neurone**
- a **relay neurone** inside the brain or spinal cord
- a **motor neurone** (see pages 69–70)

Reflex arcs are simple for good reason: it makes them as fast as possible. Generally they are fixed, which means a particular stimulus always produces the same response. This is because they do not involve the conscious parts of the brain. The two main functions of reflex arcs are to minimise damage to the body and to maintain posture by making constant, minute adjustments to muscle tone so that we remain balanced. An example of a three-neurone reflex is the knee-jerk reflex, which is important in helping to maintain posture.

Revision activity

Draw a flow diagram of the withdrawal-from-heat reflex arc. Start with 'finger touches hot plate' and end with 'hand is snatched away'.

Control of heart rate

The autonomic nervous system

Revised

Autonomic means 'self-governing' and the **autonomic nervous system (ANS)** is two sets of nerves that control many homeostatic functions without involving conscious thought. The two sets of nerves are **sympathetic** and **parasympathetic**.

Generally, impulses down sympathetic nerves prepare the body for action, in much the same way as adrenaline. The parasympathetic nerves act antagonistically to the sympathetic nerves, bringing the body back to normal and generally having a calming effect. Sympathetic nerves release a **neurotransmitter** called **noradrenaline** from their **synapses**, which is similar to the hormone adrenaline. Parasympathetic synapses release **acetylcholine (ACh)** from their synapses.

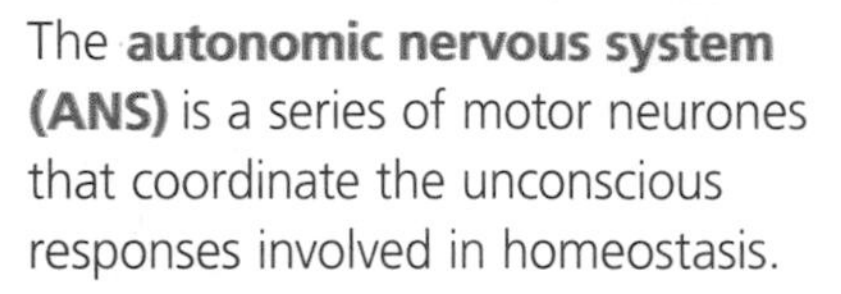

The **autonomic nervous system (ANS)** is a series of motor neurones that coordinate the unconscious responses involved in homeostasis.

A **neurotransmitter** is a substance that transmits information across the synaptic cleft between two neurones.

Synapses are junctions between nerve cells.

Acetylcholine (ACh) is a common neurotransmitter responsible for the transmission of a nerve impulse across a synapse.

Chemoreceptors and baroreceptors

Revised

An example of the ANS in action is in the control of heart rate. The rate of heartbeat is controlled by an area of the brain called the **cardioregulatory centre**, which is located in the **medulla oblongata** in the hindbrain (Figure 6.4). Within the cardioregulatory centre there is an **acceleratory centre** and a separate **inhibitory centre**. In order to match the heart rate to the demands of the body, the cardioregulatory centre gathers information from two types of receptor (Figure 6.5):

- **chemoreceptors** — chemical receptors that detect carbon dioxide concentration in the blood. These receptors are found in two places: the **carotid bodies** in the walls of the carotid arteries in the neck, and the **aortic body** in the aorta, just above the heart
- **baroreceptors** — pressure receptors in the wall of the carotid sinus, a small swelling in the carotid artery. The higher the blood pressure, the more the baroreceptors send impulses to the cardioregulatory centre. This is a sort of fail-safe mechanism that prevents blood pressure from going too high. It can also speed up the heart if it detects that blood pressure is too low

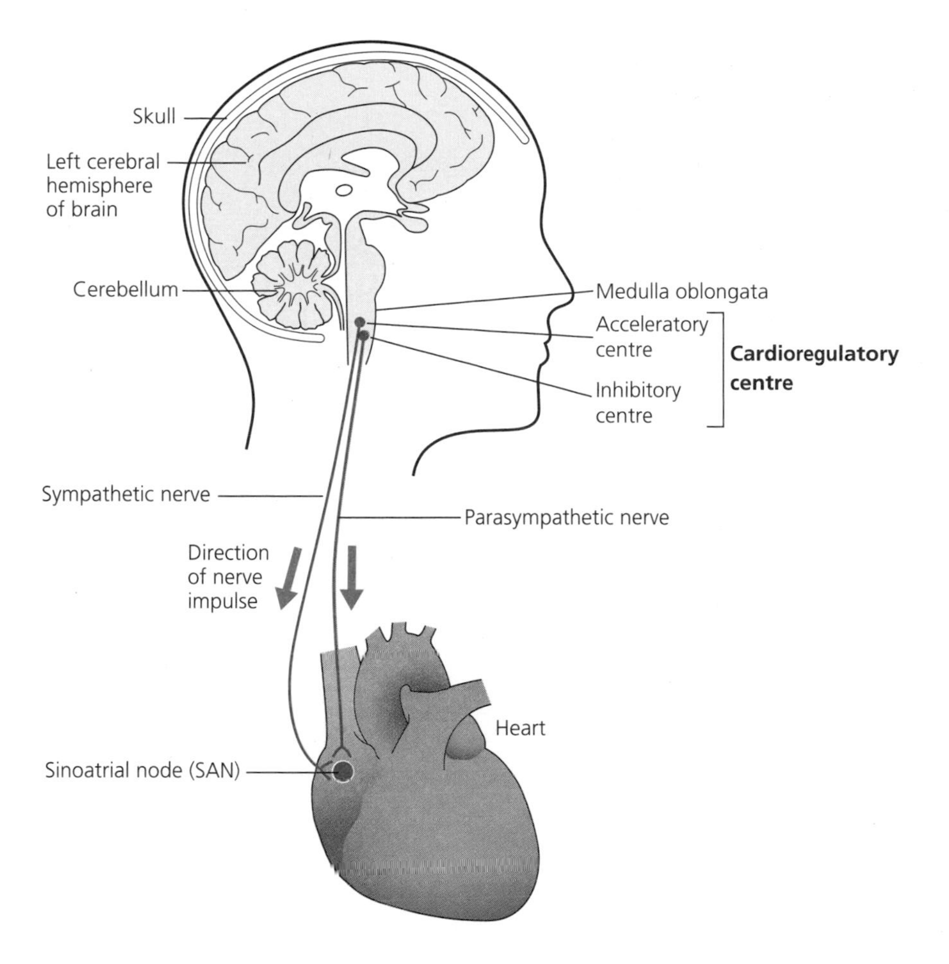

Figure 6.4 The cardioregulatory centre and the nerve supply to the heart

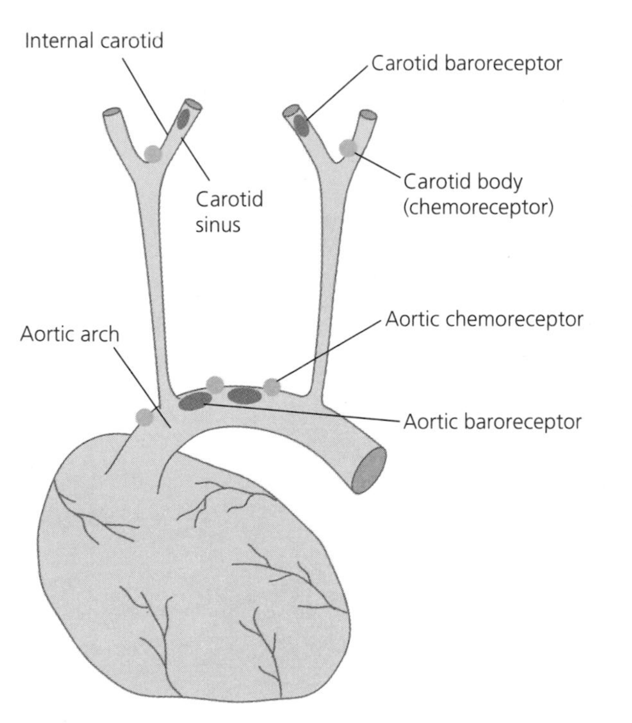

Figure 6.5 The location of chemoreceptors and baroreceptors in the blood vessels above the heart

Nerve supply to the heart

Revised

There are two nerves that pass from the cardioregulatory centre to the **sinoatrial node (SAN)** in the heart:

- the sympathetic or accelerator nerve
- the parasympathetic or decelerator nerve

The SAN sets a regular beat, but it is modified by these two nerves. When the chemoreceptors detect an increase in carbon dioxide levels, impulses pass to the cardioregulatory centre which responds by sending impulses down the sympathetic nerve. The synapse releases noradrenaline onto the SAN and the heart rate speeds up:

Exercise begins → carbon dioxide levels rise → these are detected by chemoreceptors → sensory impulses pass to the cardioregulatory centre → impulses are generated in the acceleratory centre → impulses pass down the sympathetic nerve to the SAN in the heart → heart rate speeds up

In practice, the control of heart rate is more complex and fine-tuned as a result of the balance of impulses down both nerves. There is also an element of anticipation so that exercise stimulates an increase in heart rate and breathing before an increase in carbon dioxide is detected.

Revision activity

Draw a flow diagram to summarise how heart rate slows down after exercise.

Receptors

A Pacinian corpuscle

Revised

A **Pacinian corpuscle** is one example of the different receptors found in mammalian skin. Its function is to detect heavy pressure and vibration.

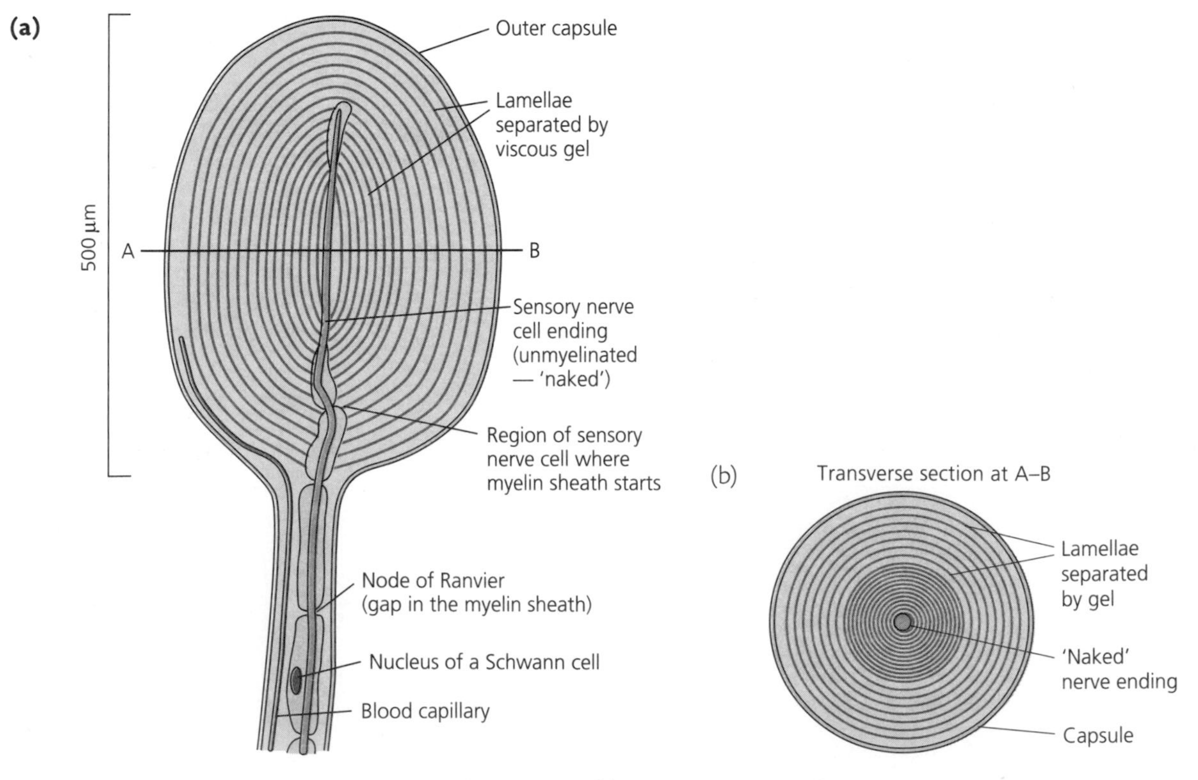

Figure 6.6 (a) A Pacinian corpuscle in longitudinal section (b) A Pacinian corpuscle in transverse section

Each corpuscle is about 1 mm deep and consists of 20–60 lamellae (layers of cell membrane) each separated by a gel, so its structure resembles an onion. At the centre is a single receptor cell, often called a nerve ending. In the membrane of the receptor are proteins called **stretch-mediated sodium channels**.

How a Pacinian corpuscle works:

1. Pressure causes the proteins to change shape and become more permeable to sodium ions.
2. Sodium ions rapidly diffuse in, creating a **generator potential**.
3. If the generator potential reaches a **threshold value**, an **action potential** is generated in the sensory nerve and we get the sensation of pressure.
4. The receptor adapts very quickly, so it does not generate prolonged impulses. The gel quickly redistributes to even out the pressure, so the membrane proteins return to their normal permeability.

The vital points about this receptor (and most receptors) are:

- it only responds to one type of stimulus
- it stops responding to a prolonged stimulus
- stronger stimuli generate more frequent impulses

Now test yourself Tested

7 Suggest the advantage of receptors adapting so that they do not respond to prolonged stimuli.

Answer on p. 111

The retina

Revised

The retina is the layer of light-sensitive cells — **rods** and **cones** — that gather visual information and channel it down the **optic nerve** to the brain, where an image is formed. There are two key aspects to the retina:

- **sensitivity** — the ability to see in dim light
- **visual acuity** — the ability to see detail

The rods are more sensitive because several of them **converge** into one sensory neurone, allowing them to **summate** (Figure 6.7). All the rods supplying the neurone can contribute to reaching the threshold. The cones have a higher visual acuity because there is much less convergence — at the centre of the **fovea** each cone has its own sensory neurone. Consequently, the cones send more information to the brain per unit area of retina.

The **fovea** is the part of the retina that has no rod cells but large numbers of cone cells. Therefore, it is the region of highest visual acuity. In other words, it is the part of the retina that enables the greatest degree of detail to be seen.

Examiner's tip

Visual acuity can be thought of as resolution — the ability to distinguish between two close objects.

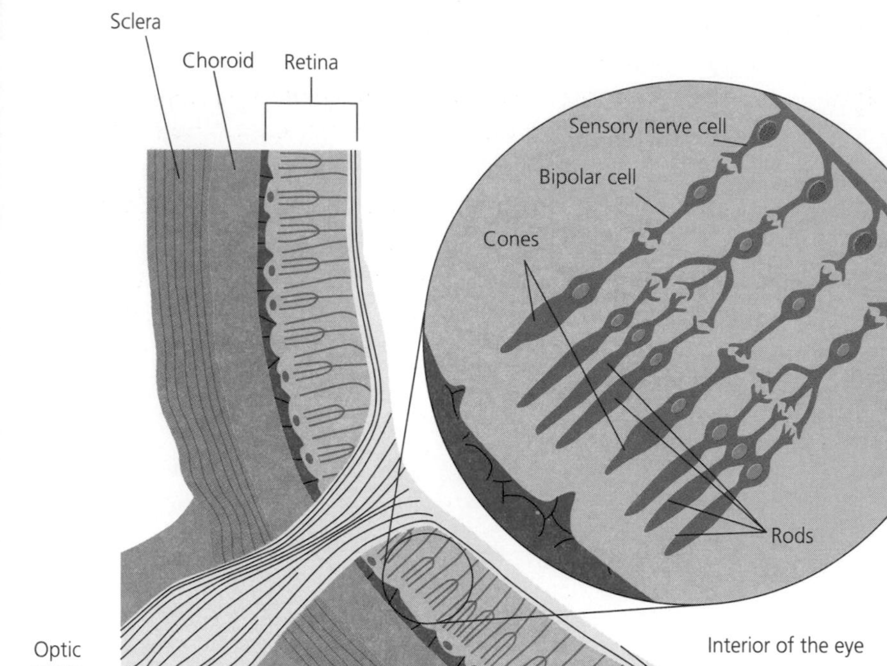

Figure 6.7 The distribution of rods and cones in the retina. Note that several rods (shown here as three) converge into one sensory nerve cell, whereas the cones do not converge

Table 6.1 Comparing rods and cones

Feature	Rods	Cones
Distribution	Periphery of retina	Concentrated at the fovea. A few in the periphery
Sensitive to	Dim light	Bright light
Amount of retinal convergence	Many rods converge into one neurone	Each cone has its own neurone
Overall function	Vision in poor light	Vision in colour and detail in good light

Now test yourself

8 Explain what is meant by the term *visual acuity*.

Answer on p. 111

Tested

Exam practice

1 The following graph shows the distribution of rod and cone cells across the retina of the human eye.

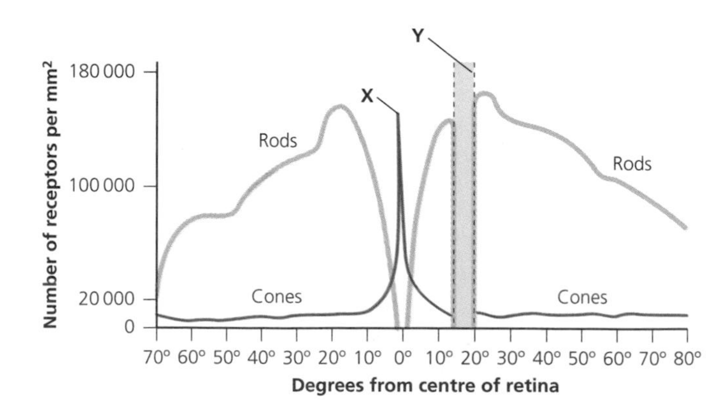

(a) Explain why rods have a greater sensitivity than cones. [2]

(b) We can only see in detail, such as the exact word you are reading on this page, because an image of the word falls on point X. Explain why point X allows such a high visual acuity. [2]

(c) Explain why, when an image is formed on area Y of the retina, nothing is perceived by the brain. [2]

(d) When looking at a particularly dim star, it appears clearer if we do not look straight at it. Use your knowledge of rod cells to explain why. [2]

Answers and quick quizzes online

Online

Examiner's summary

By the end of this chapter you should be able to understand:

- ✔ Tropisms are simple growth responses in plants, controlled by indoleacetic acid (IAA).
- ✔ Taxes and kineses are simple responses that allow mobile organisms to find favourable conditions.
- ✔ A simple reflex arc is the simplest coordinated response and involves three neurones.
- ✔ Heart rate is controlled by chemoreceptors and baroreceptors, the autonomic nervous system and effectors.
- ✔ The Pacinian corpuscle is an example of a receptor.
- ✔ In the retina, the differences in sensitivity and visual acuity can explained by the distribution of rods and cones and the connections they make in the optic nerve.

7 Nerves and muscles

Principles

Nerves and hormones

Revised

There are two ways to communicate:

- **Nervous communication** consists of **neurones** passing **electrical impulses** along their length. Neurones pass directly to **target cells**. At the ends of the neurones are synapses, which stimulate the target cells by secreting chemical **neurotransmitters** directly onto them.
- **Hormonal communication** consists of the **endocrine system** secreting **hormones** directly into the bloodstream. They pass to all parts of the body, but stimulate their target cells in a lock and key fashion: the target cells have specific **hormone receptor proteins** on their surface.

Hormones are chemical signalling messengers that are released directly into the bloodstream.

The key differences are summarised in Table 7.1.

Table 7.1 Comparing nervous and hormonal communication

Feature	Nerves	Hormones
Nature	Electrical/ionic impulses passing down neurones	Chemicals travelling in the bloodstream
Precision	Very precise — impulses pass directly to target cells	Widespread because hormones reach all parts of the body, but also precise because only target cells have receptor proteins for that specific hormone
Speed of action	Very rapid	Usually slow, although adrenaline is fast
Duration of action	Very short lived	Usually prolonged
Modulation — how can we distinguish between large and small stimuli?	Frequency: greater stimulus = more impulses per second	Amplitude: greater stimulus = greater concentration of hormone

Now test yourself

Tested

1 Explain how a hormone produced in the brain can affect just the ovaries and nowhere else.

Answer on p. 111

Histamine and prostaglandins

Revised

There are several substances, known as **local chemical mediators**, that are made by cells and have an effect only on the surrounding cells. Two notable examples are:

- **histamine**, which is the chemical largely responsible for inflammation. It is made by white cells called **mast cells** in response to cell damage or bacterial infection. Histamine affects cells locally; it makes blood capillaries more permeable so that white blood cells and fluid accumulates in the affected area

- **prostaglandins**, which are a group of substances with a variety of localised effects, including inflammation. They were first found in the prostate gland, hence the name, but they are secreted by most organs in the body. Their effects include stimulating contraction of smooth muscle (such as the uterus) and controlling platelets during blood clotting.

Examiner's tip

It is unlikely that an exam question will ask you to know details about the action of local chemical mediators. It is more likely that you will be asked about the difference between them and hormones.

Now test yourself

Tested

2 Explain the difference between a local chemical mediator and a hormone.

Answer on p. 111

Nerve impulses

Motor neurones

Revised

A **motor neurone** is a nerve cell that is specialised to transmit impulses. Its key features are:

- a cell body that contains the nucleus and other organelles
- an elongated axon that carries impulses away from the cell body
- one or more dendrites that take impulses towards the cell body

A **motor neurone** is a cell that carries nerve impulses away from the central nervous system to an effector, such as a gland or muscle.

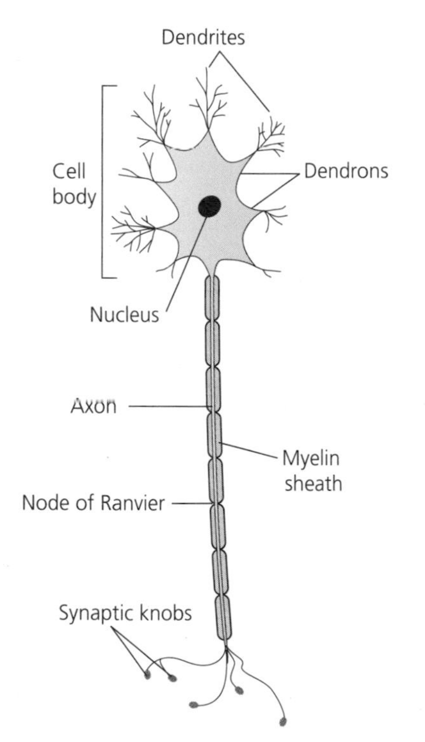

Figure 7.1 The basic structure of a motor neurone

Typical mistake

Nerve impulses are not messages. 'Nerve impulse' and 'action potential' are good A-level terms, but don't call them messages or signals.

Now test yourself

3 Where in the neurone is the nucleus?

4 What is the difference between an axon and a dendrite?

Answers on p. 111

Tested

Many neurones are **myelinated**, which means that their axon is protected by a fatty **myelin sheath** made from several layers of cell membrane. The sections are made by individual **Schwann cells** that wrap around the axon many times. In between the sections of myelin are nodes of Ranvier where the axon membrane is exposed. The role of the myelin sheath is to:

A **myelin sheath** is a non-conducting fatty layer around the axon.

- insulate the axon
- protect the axon
- speed up the transmission the nerve impulses (see pages 71–72)

Resting potential

Revised

The **resting potential** is a state of readiness in a neurone. In a resting cell, there is a potential difference across the membrane of about –70 mV. Once a resting potential is established, the neurone is ready to transmit an impulse. A resting potential is the result of two processes occurring together:

- **active transport** — the sodium–potassium pump is a protein responsible for the active transport of positive ions. For each ATP molecule split, 3 **sodium** (Na^+) ions are pumped out of the axon and 2 **potassium** (K^+) ions are pumped in (Figure 7.2)
- **unequal facilitated diffusion** — there are sodium ion channels and potassium ion channels. Both are gated so that, by changing their shape, they can open up and allow the ions to diffuse freely. The key difference is that the potassium channels are more 'leaky' than the sodium channels, so the potassium diffuses out faster than the sodium diffuses in

The **resting potential** is the potential difference across the axon membrane of a nerve cell when the cell is at rest.

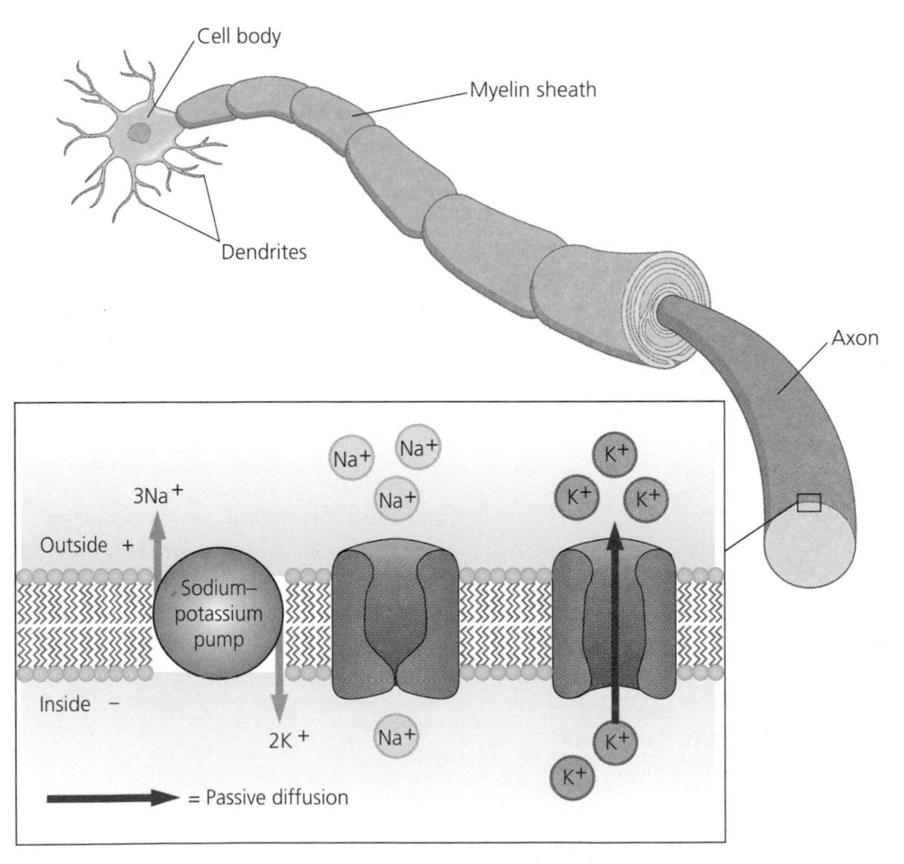

Figure 7.2 A resting potential across the axon membrane

The overall result is a high concentration of positive ions outside the axon, giving the area a positive charge relative to the inside. This is a resting potential. There are also negative ions inside and outside the axon membrane, the most common being chloride ions and negatively charged proteins. However, it is the movement of positive ions that establishes a resting potential and brings about an **action potential**, so we focus on those.

Revision activity

Look at an animation of the mechanisms of resting and action potential — there are lots available.

Now test yourself

5 What are the two key processes that result in the creation of a resting potential?

6 What is the value of the average resting potential in mV (millivolts)?

Answers on p. 111

Generation of an action potential

Revised

An action potential is a **nerve impulse**. It is a rapid reversal of a resting potential that spreads rapidly along the axon. It is started by one simple action: the gated sodium ion channels open for a fraction of a second so that sodium ions diffuse rapidly into the axon (Figure 7.3). For about a millisecond, the resting potential is reversed in one area of the axon so that the inside becomes positively charged (with respect to the outside). This reversal spreads quickly along the axon while the original area recovers and establishes a resting potential again.

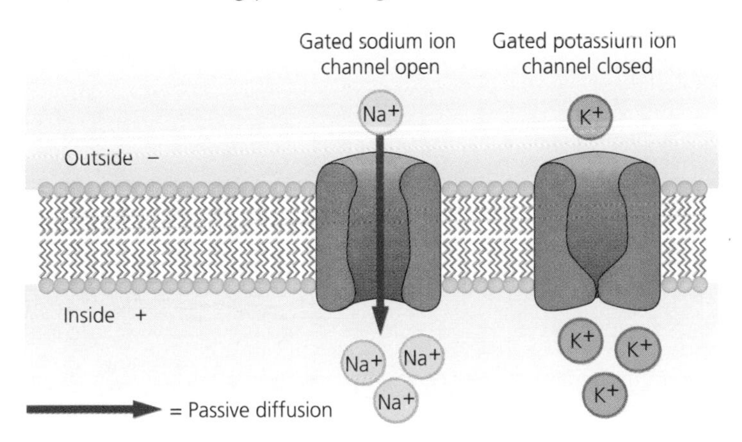

Figure 7.3 An action potential is initiated by the opening of gated sodium ion channels

The key events are:

1. **Depolarisation** — the gated sodium ion channels open, allowing sodium ions to diffuse in and reverse the resting potential.
2. **Repolaristion** — the gated sodium ion channels close, preventing further diffusion inwards. The gated potassium ion channels open fully, allowing positively charged potassium ions to diffuse out rapidly.
3. **Hyperpolarisation** — an 'undershoot' results from the gated potassium ion channels still being open while the active transport mechanism begins to re-establish a resting potential by pumping sodium ions out.
4. The gated potassium ion channels return to their normal permeability and a resting potential is re-established.

These events, which occur over a short period of time at one place in the axon, are shown in Figure 7.4.

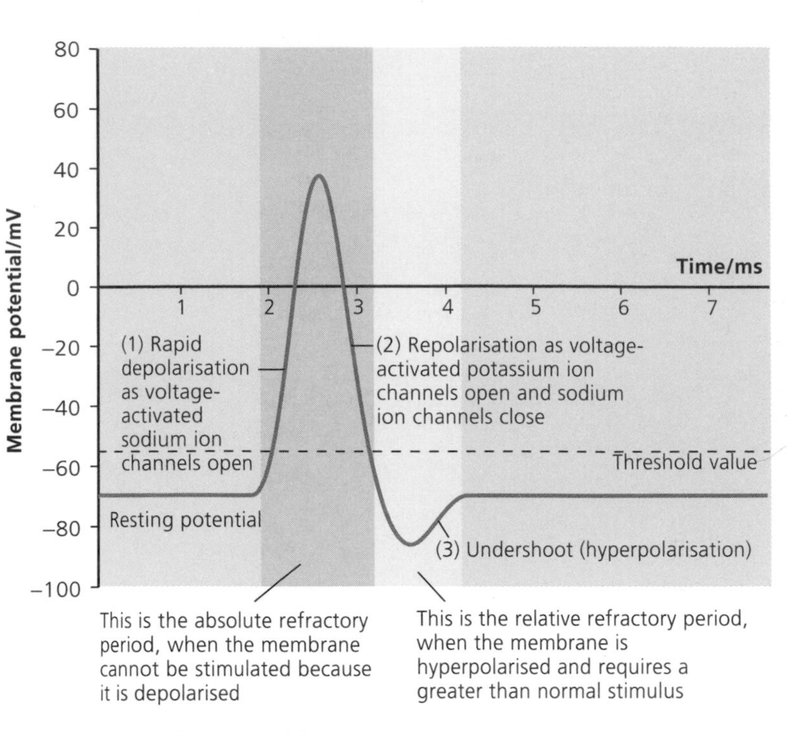

Figure 7.4 An action potential trace

An action potential is **self-propagating**, which means that the depolarisation of one region of an axon will immediately cause the depolarisation of the next region. The gated sodium and potassium ion channels in the axon membrane are 'voltage-gated', which means that their shape, and therefore their permeability, depends on the voltage (charge) across the membrane.

The all-or-nothing principle

Revised

If the generator potential is large enough, and its **threshold value** is reached, an action potential is generated. There either *is* an action potential or there *is not*. If the generator potential is not large enough, there is no action potential. This is the **all-or-nothing principle**. The brain makes sense of the incoming sensory information. It knows:

- the origin of the incoming impulses — if the stretch receptors in your bladder send impulses to the brain, you know what it means and what you need to do
- the frequency of the incoming impulses — the more intense the stimulus, the more frequent the impulses. It is not possible to have a small action potential for a small stimulus and a larger one for a large stimulus

The **all-or-nothing principle** describes a nerve impulse. When an action potential is produced in a nerve cell, it is always the same size. It does not matter how big the initial stimulus, the action potential will always involve the same change in potential difference across the cell surface membrane.

The refractory period

Revised

After an impulse has passed, there is a short period of time when it is impossible to initiate a new action potential. This is the **absolute refractory period**. This is followed by a brief period when it is possible to generate an impulse but the threshold value is greater, so the stimulus must be of greater intensity than normal. This is the **relative refractory period**.

The **relative refractory period** is the short recovery period that occurs immediately after the passage of a nerve impulse along the axon of a nerve cell.

The refractory period is important because it separates nerve impulses from each other. Both phases of the refractory period are shown in Figure 7.4.

Factors affecting the speed of conductance

Revised

Several factors affect the speed of conductance of a nerve impulse:

- The conduction of impulses along myelinated neurones is faster than in non-myelinated neurones because the impulse jumps from one node to the next. This is called **saltatory conduction**. Myelin prevents depolarisation from happening, so it only happens at the nodes where the axon membrane is exposed.
- The wider the **axon diameter**, the faster the transmission. Generally, this is only important in invertebrates, which have never evolved myelin.
- The higher the **temperature**, the faster the transmission because the movement of ions is faster.

The speed of transmission from one part of the body to another is also affected by the number of synapses in any given pathway. Synapses slow transmission down, which is why there are as few as possible in a reflex arc (see pages 61–62).

Synaptic transmission

Structure and function

Revised

A **synapse** is a junction between two neurones or between a neuron and a muscle, in which case it is called a **neuromuscular junction**. Synapses are found at the end of axons (Figure 7.5).

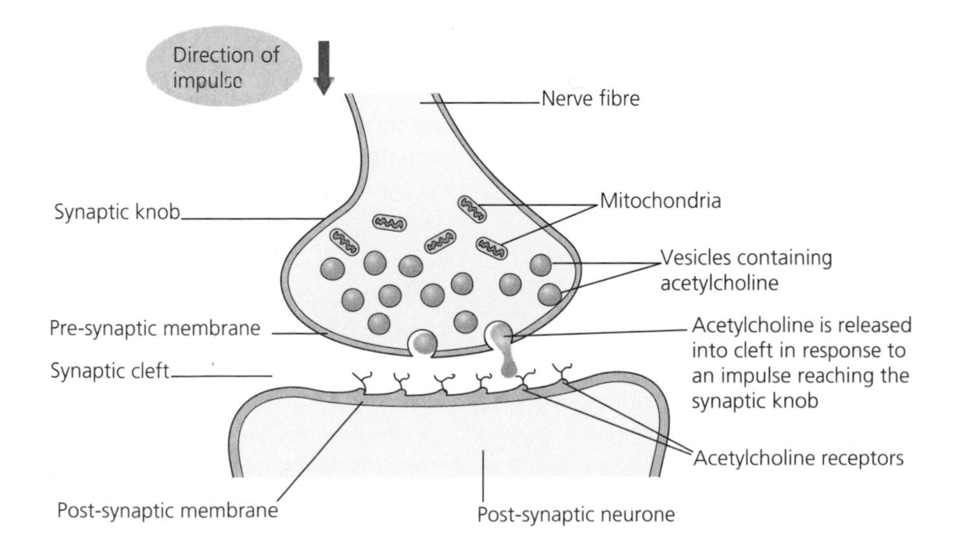

Figure 7.5 The structure of a synapse

The events of synaptic **transmission** are shown in Figure 7.6. Immediately after transmission, an enzyme in the **synaptic cleft** breaks down the neurotransmitter. The products of this breakdown are reabsorbed into the

> The **synaptic cleft** is the small gap between the pre-synaptic neurone and the post-synaptic neurone at a synapse.

pre-synaptic membrane to be resynthesised into the active transmitter again. In many synapses outside the central nervous system, the neurotransmitter is **acetylcholine (ACh)**. These synapses are said to be **cholinergic synapses**.

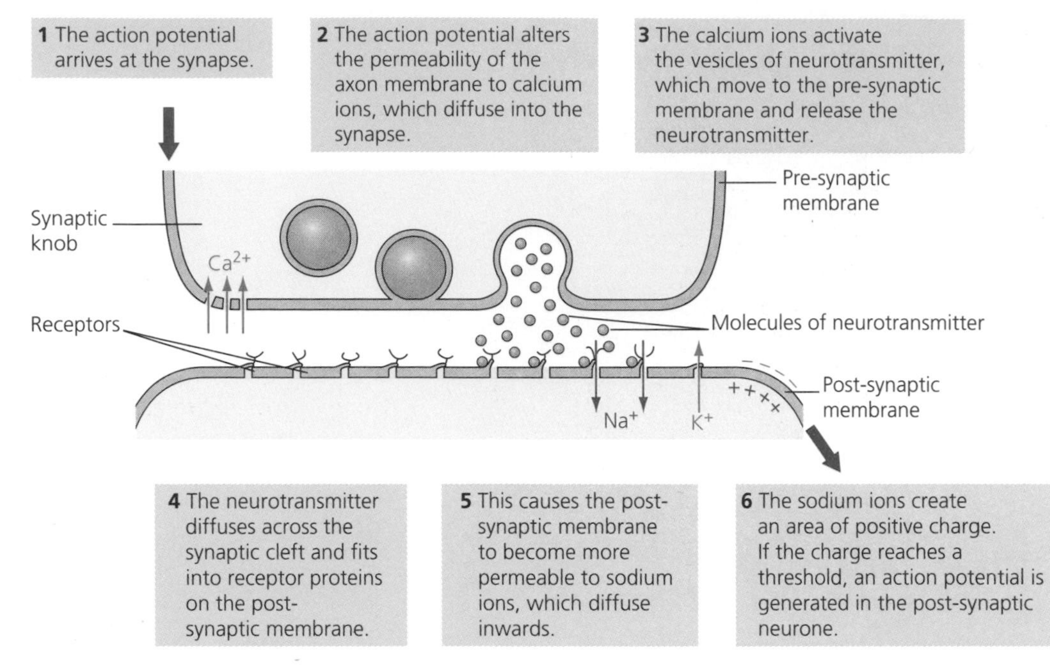

Figure 7.6 Synaptic transmission

Now test yourself — Tested

7 Fill in the gaps in the following flow diagram to describe synaptic transmission.
The action potential arrives at the synapse → ______________ ions flow into the pre-synaptic membrane → molecules of neurotransmitter ______________ across the synaptic cleft and fit into specific ______________ ______________ → The ______________ of the post-synaptic membrane changes → ______________ ions flow in, causing a positive charge to build up inside the ______________ synaptic membrane → If ______________ is reached, an action potential is created in the post-synaptic neurone.

8 Suggest two reasons for the presence of mitochondria in the synaptic knob.

Answers on p. 111

Key features of synapses

Revised

Synapses are **unidirectional**, which means an impulse can only pass one way. This is because the neurotransmitter can only be made and released from the pre-synaptic side and the receptor proteins are found only on the post-synaptic side.

Some synapses are **inhibitory**. It is just as important to be able to switch synapses off, so that precise pathways can be selected. An inhibitory synapse makes it less likely that an impulse is generated in the post-synaptic membrane. As shown in Figure 7.7, neurones A and B are excitatory synapses, but neurone C is an inhibitory synapse.

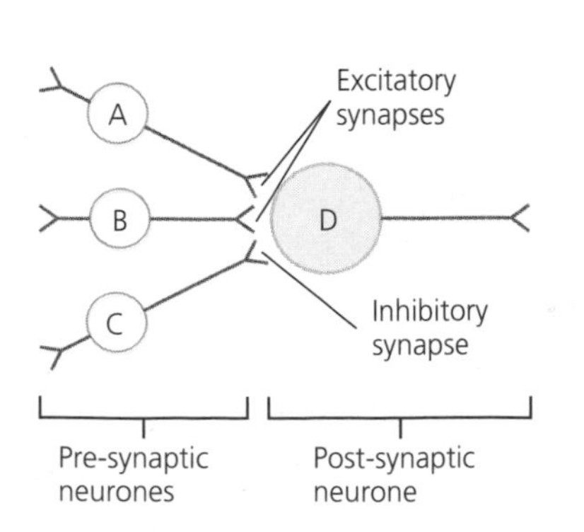

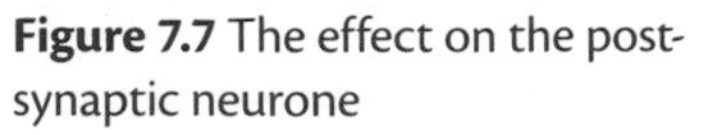
Figure 7.7 The effect on the post-synaptic neurone

Synapses can also **summate**, which means they combine to generate an impulse. There are two types of summation:

- **temporal summation** (separated in time) — impulses are transmitted down two or more separate synapses at the same time
- **spatial summation** (separated in space) — impulses are transmitted down the same synapse in quick succession

Table 7.1 Summation at a synapse

Impulses down pre-synaptic neurones	Impulse generated in D?	Explanation
One impulse down A	No	Threshold is not reached
One impulse down B	No	Threshold is not reached
Two impulses down A in quick succession	Yes	Temporal summation — both impulses combine to reach the threshold
Two impulses down A and B at the same time	Yes	Spatial summation — both impulses combine to reach the threshold
Three impulses down A, B and C at the same time	No	The threshold is not reached because neurone C is inhibitory, so it cancels out one excitatory impulse

Now test yourself

9 Suggest why it is important that synapses are unidirectional.

Answer on p. 111

Tested

The effect of drugs

Revised

Many **drugs** (legal or otherwise) exert their effect by interfering with normal synaptic transmission. For example, some drugs:

- block the calcium channels
- prevent synthesis of the neurotransmitter
- fit into the post-synaptic receptors
- prevent re-uptake of the neurotransmitter
- block the enzyme that breaks down the neurotransmitter

Examiner's tip

You don't need to learn the action of any specific drugs. Exam questions will test your understanding of synapses by telling you how specific (and probably obscure) drugs work.

Now test yourself

Tested

10 For each of the effects of drugs listed above, predict the effect on synaptic transmission.

Answer on p. 111

Skeletal muscles

Types and function of skeletal muscle

Revised

Skeletal muscles produce movement and maintain posture. They are attached to bones by tendons and are generally under conscious control. Muscle is a specialised tissue that can do just one thing — contract — and so skeletal muscles generally work **antagonistically** in pairs or groups. One contracts while the other relaxes.

There are two other types of muscle: **smooth muscle**, which is found in tubular organs such as the gut, reproductive system and blood vessels, and **cardiac muscle**, which is found only in the heart.

The sliding filament theory of muscle contraction

The structure of skeletal muscle

Revised

Skeletal muscle is made from many small cylindrical fibres called **myofibrils**. Along their length is an alternating pattern of light and dark bands caused by two key interlocking proteins, **actin** and **myosin**. Each repeated unit of actin and myosin is called a **sarcomere**. Muscular contraction is caused when myosin pulls itself into the actin, shortening all the sarcomeres and therefore the muscle as a whole.

In a resting muscle, a thin fibrous protein called **tropomyosin** covers the sites where actin binds to myosin. **Calcium ions** are essential for contraction, but they are stored outside the myofibril, in the **sarcoplasmic reticulum**. Figure 7.8 explains the theory.

Myofibrils are the small fibres that are arranged parallel to each other in a skeletal muscle fibre. 'Myo' means muscle.

Sarcoplasmic reticulum is specialised endoplasmic reticulum.

(a) A section through a muscle fibre showing the myofibrils. The sarcoplasmic reticulum (in blue) stores the calcium ions essential in contraction

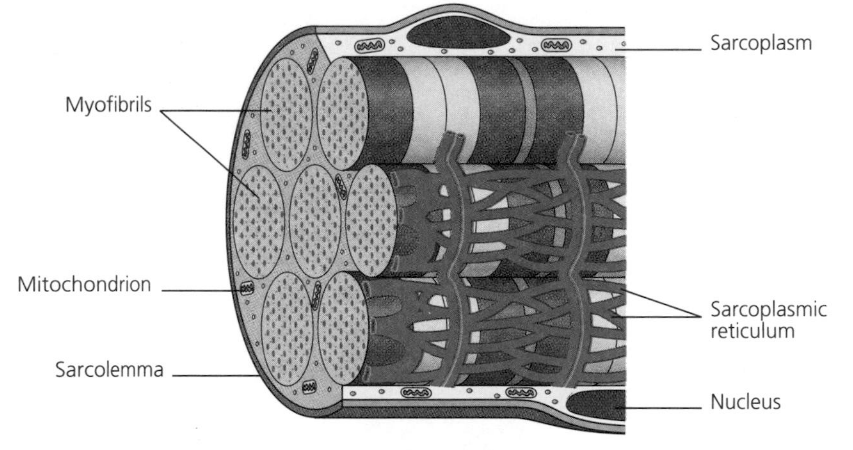

(b) The repeated pattern of actin and myosin fibres in two sarcomeres

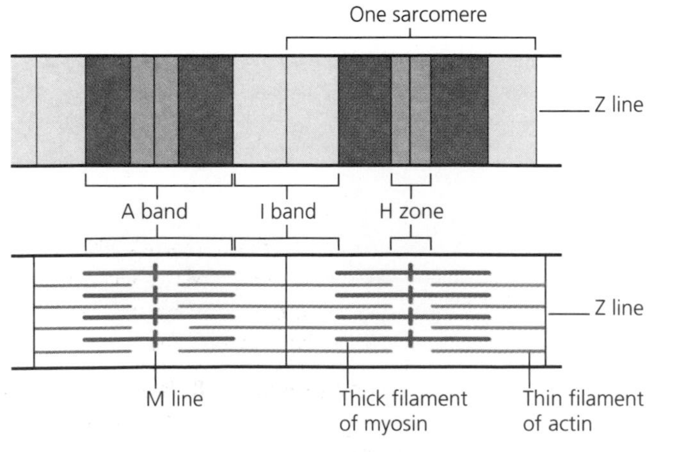

(c) The same sarcomeres after contraction

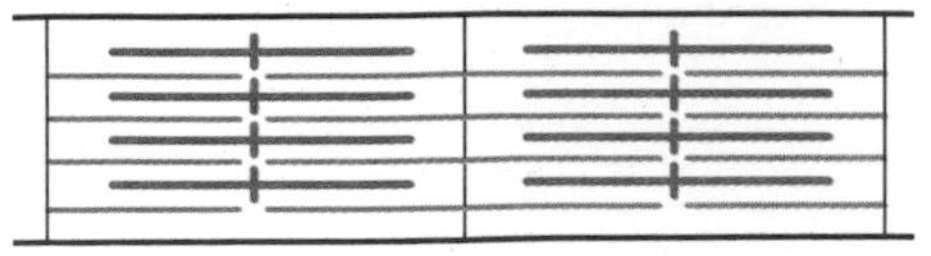

Figure 7.8 The sliding filament theory of muscle contraction

Examiner's tip

Don't worry about learning the different bands — A band, I band etc. Exam questions will ask you about the mechanism of contraction and how the banding pattern of sarcomere changes.

Muscles as effectors

How does a muscle contract?

Revised

When a muscle fibre is stimulated by nerve impulses, the following steps occur that result in contraction of the muscle.

An **action potential** arrives down a motor neurone, reaching the **neuromuscular junction**. In a similar mechanism to synapses, the neurotransmitter **acetylcholine (ACh)** is released onto the **motor end plate**. The neurotransmitter causes a wave of **depolarisation** to spread along and around the myofibrils in the sarcoplasmic reticulum and T tubules. **Calcium ions** are released from the sarcoplasmic reticulum and into the myofibril.

The calcium ions bind to **troponin**, a small globular protein that binds to **tropomyosin** and moves it away from the actin–myosin binding site. The **myosin heads** can then bind to the actin molecules and form the **actinomyosin bridges**. The **ATP** attached to the myosin head splits. This releases the energy that makes the myosin head bend, pulling the actin along. Another ATP molecule attaches to the myosin head and splits. The energy is used to detach the myosin head, change the angle of the head ('re-cock') and re-attach further along the actin. The process repeats and the myosin pulls itself over the actin, shortening the sarcomere.

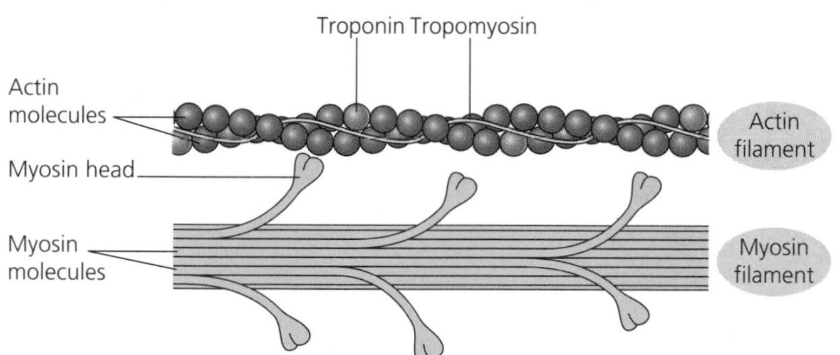

Figure 7.9 The arrangement of the four major proteins in muscular contraction

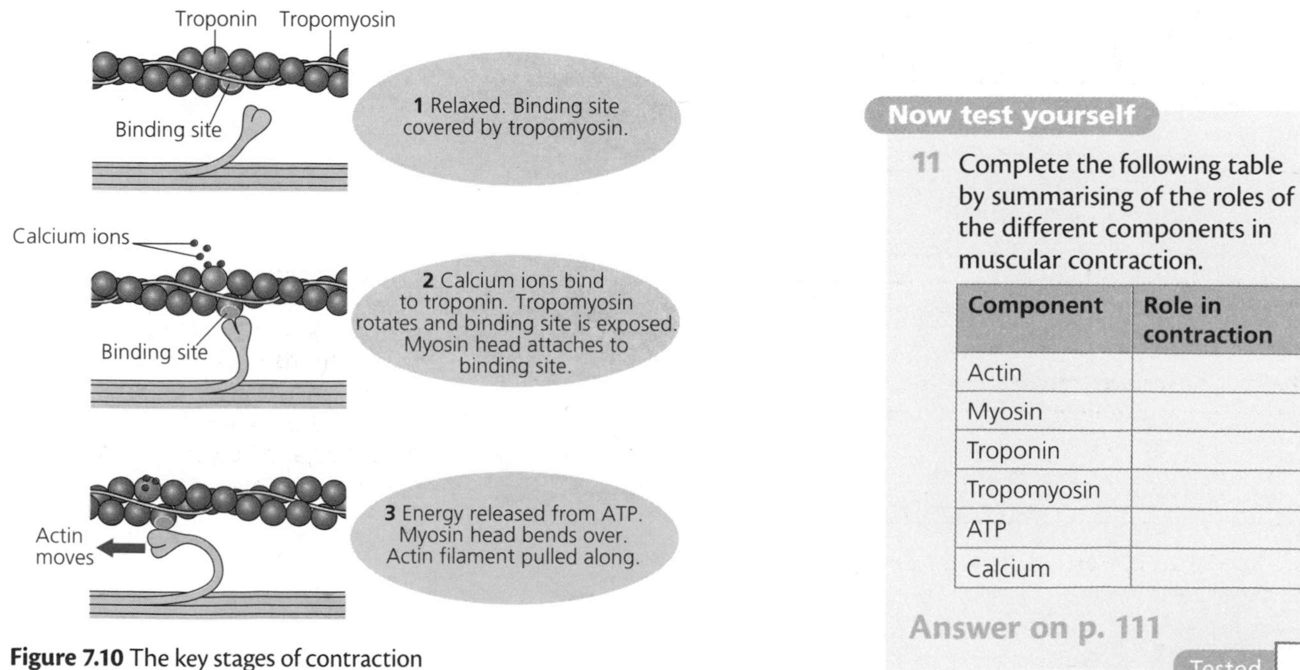

Figure 7.10 The key stages of contraction

Now test yourself

11 Complete the following table by summarising of the roles of the different components in muscular contraction.

Component	Role in contraction
Actin	
Myosin	
Troponin	
Tropomyosin	
ATP	
Calcium	

Answer on p. 111

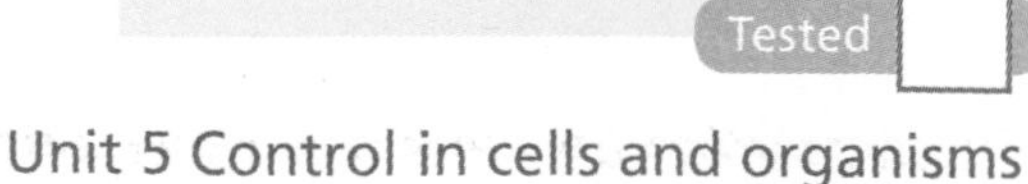

Sources of ATP

Revised

Muscles cannot move without ATP and prolonged exercise needs a continuous supply. ATP can come from three sources, depending on the timescale:

- the ATP already present in the muscles, which has accumulated during periods of rest. However, this only lasts for a few seconds during vigorous exercise. There is another chemical, **phosphocreatine (PC)**, which can be hydrolysed instantly to resynthesise ATP. This ATP/PC system provides ATP for about the first 10 seconds of vigorous exercise
- ATP from **glycolysis** — the first stage of respiration. This is anaerobic respiration that yields only 2 ATP molecules per glucose molecule and comes with the problem of lactate build-up. However, it is quick and bridges the gap between the first and last sources, i.e. between 10 seconds and about 1 minute
- **full aerobic respiration** — the reactions that take place in the electron transport chain (oxidative phosphorylation) can provide over 34 ATP molecules per glucose molecule and there is no lactate build-up. However, it is slow and can only provide ATP at a certain rate. This is why you cannot sprint for long distances. The first two sources can provide ATP for contraction at full power, but the aerobic system can only provide ATP for about 60–70% of the maximum

Phosphocreatine (PC) is a substance that replaces and provides a supply of ATP for immediate use. PC can be replenished from the ATP produced in respiration.

Fast and slow skeletal muscle fibres

Revised

There are two types of skeletal muscle fibres, **fast skeletal muscle fibres** and **slow skeletal muscle fibres**, named according to their speed of contraction. Most people are born with approximately equal numbers of the two fibre types, although training can alter the balance. Others are born with a lot of slow or fast muscle fibres, which makes them natural athletes at either endurance or strength events.

Table 7.2 Comparison of slow and fast skeletal muscle fibres

Feature	Fast skeletal muscle fibres	Slow skeletal muscle fibres
Speed of contraction	Fast, powerful	Slow
Speed of fatigue	Rapid; lactate accumulates and an oxygen debt builds up	Slow
Main source of ATP	Glycolysis (anaerobic respiration)	Electron transport chain (aerobic respiration)
Structure of fibres	Thicker in diameter; pale in colour	Thin cross-section; red in colour due to dense mitochondria; lots of blood capillaries and myoglobin
Location	Arms and legs	Postural

Exam practice

1 The following diagram shows a neuromuscular junction and a myofibril.

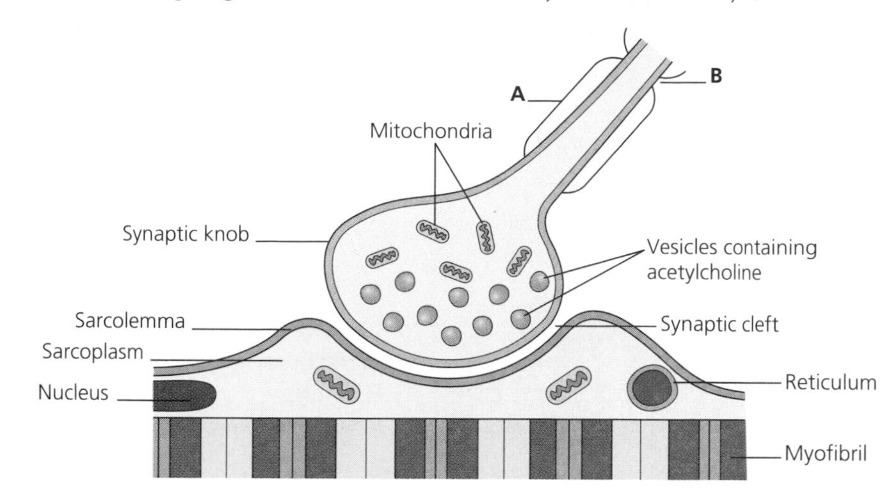

(a) Name the features labelled **A** and **B**. [1]

(b) Explain the advantage of the synaptic cleft being small. [1]

(c) The toxin curare was discovered in South America, where it was used to tip poison darts. It works by competing with acetylcholine for the receptors on the post-synaptic membrane. Predict the effects of curare on muscular contraction. Explain your answer. [3]

2 There are two types of muscle fibre, fast and slow. Slow skeletal muscle fibres are generally smaller in diameter and deep-red in colour due to the presence of many mitochondria and the oxygen storage molecule myoglobin. Suggest why slow skeletal muscle fibres are:

(a) small in diameter [2]

(b) rich in mitochondria [2]

(c) rich in myoglobin [2]

Answers and quick quizzes online

Online

Examiner's summary

By the end of this chapter you should be able to understand:

- ✔ Nervous control results in rapid, short-lived and localised responses.
- ✔ Hormonal control results in slow, long-lasting and widespread responses.
- ✔ Histamine and prostaglandins are local chemical mediators released by some mammalian cells that affect only cells in their immediate vicinity.
- ✔ The basic structure of a motor neurone.
- ✔ How a resting potential is established as a result of unequal ion distribution.
- ✔ The ionic events that lead to the action potential.
- ✔ The all-or-nothing principle.
- ✔ The nature and importance of the refractory period.
- ✔ Factors affecting the speed of conductance.
- ✔ The structure of a synapse and neuromuscular junction.
- ✔ The events involved in transmission across a synapse and neuromuscular junction.
- ✔ How to interpret the action of a given drug.
- ✔ The gross and microscopic structure of skeletal muscle, especially myofibrils.
- ✔ The roles of actin, myosin, tropomyosin, troponin, calcium ions and ATP in muscle contraction.
- ✔ The role of ATP and phosphocreatine in providing energy during contraction.
- ✔ The structure, location and general properties of slow and fast skeletal muscle fibres.

8 Homeostasis and hormones

Principles of homeostasis

Maintaining a constant internal environment

Revised

The word **homeostasis** means 'steady state', but the conditions within the body of a mammal are controlled within certain limits rather than being kept constant. Overall, there are many homeostatic mechanisms that work together to perform a remarkably effective balancing act, such as:

- maintaining a constant **core temperature** of around 37°C
- maintaining constant **blood pH** at between 7.3 and 7.45

Both conditions need to be kept within narrow limits because of the sensitivity of enzymes. By contrast, **blood glucose concentrations** can vary widely. If levels fall too low, the body's cells become starved of vital fuel. If they rise too high, the water potential of the blood falls too low. However, as long as the levels are kept at between 4 and 11 mmol/L, there are usually no problems.

Homeostasis is the maintenance of constant internal conditions, such as temperature, pH, water potential and blood glucose concentration.

Typical mistake

Candidates tend to write that 'enzymes are denatured' as the explanation for all enzyme-related problems. However, if the temperature rises too high or the pH is not quite at the optimum level, different enzymes will be affected to a different extent and the result is a metabolic imbalance. It takes more extreme conditions to denature most enzymes.

Principles of negative and positive feedback

Negative feedback

Revised

Negative feedback is the mechanism that keeps things stable (Figure 8.1). The key elements are:

1. A physiological level changes — for example, temperature increases or blood glucose decreases.
2. The change is detected by receptors.
3. A mechanism is activated that reverses the change.
4. The extent of the change is monitored so that when conditions return to normal, the corrective mechanism is switched off.

Negative feedback is the process by which a departure from the set point results in changes that lead to a return to the original value.

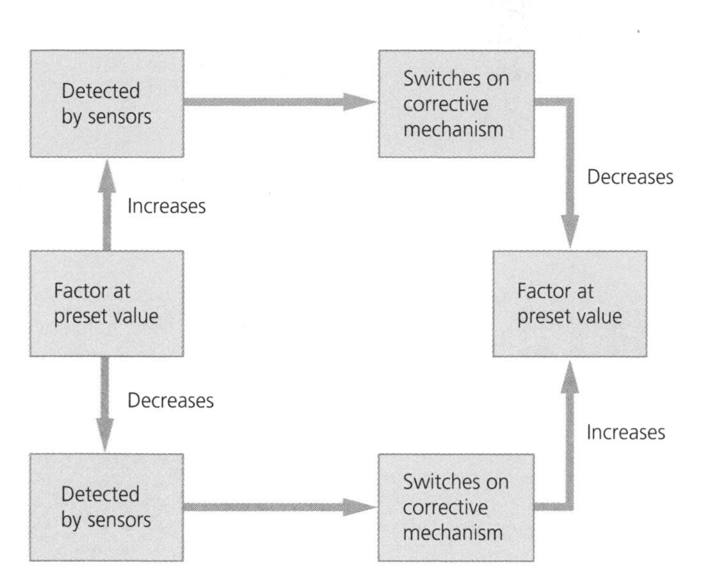

Figure 8.1 Negative feedback is often described as a 'detection-correction' mechanism as deviations from a set point are returned to the norm

Positive feedback

Revised

Positive feedback is the mechanism for change. It is often said that positive feedback is an emergency in which normal homeostatic mechanisms have broken down and this can be true (see page 83). However, there are many situations in which positive feedback brings about a necessary change, such as blood clotting or childbirth.

Positive feedback involves a departure from the set point, bringing about changes that produce further change.

Now test yourself

Tested

1 What is the key difference between positive and negative feedback?

Answer on p. 112

Temperature control

Contrasting mechanisms

Revised

Animals can be divided into two groups according to their ability to control their core temperature:

- **Endotherms** can maintain a constant core temperature regardless of the environmental temperature. They are often called warm blooded. Only mammals and birds are endotherms. They use **physiological control mechanisms** and behaviour to control body temperature.
- **Ectotherms** are animals that cannot maintain a constant core temperature. Their bodies tend to be at the same temperature as their surroundings. They can **thermoregulate** by behavioural methods, such as sunbathing, but do not have the physiological mechanisms.

Endotherms are animals that maintain their body temperature using physiological mechanisms.

Ectotherms are animals that make use of the environment to regulate their body temperature.

Examiner's tip

The term *core temperature* refers to the centre of the body, not the extremities such as hands and feet, which are often a lot colder. The specification refers to temperature control in reptiles and mammals only.

Temperature control in mammals

Revised

Temperature control is another example of the autonomic nervous system in action. The idea behind it is that **metabolism** — the chemical reactions of the body — produces **heat** as a by-product.

A constant temperature is maintained by minimising or maximising heat loss to the environment. All temperature control mechanisms need a thermostat and in mammals this is located in the **hypothalamus** in the brain. This is sensitive to the temperature of the blood that flows through it. If the core temperature rises, it is detected by hypothalamus. The **heat loss centre** in the hypothalamus initiates several responses by sending autonomic (involuntary) impulses to effectors in the skin (Figure 8.2):

- **vasodilation** (Figure 8.3) — skin arterioles dilate so that more blood can flow through the capillaries just under the surface of the skin
- **sweating** — heat from the blood is transferred to the sweat and is lost as the water evaporates

Other responses to heat include the hairs lying flat and, in the long term, a decrease in metabolic rate.

Now test yourself

2 Distinguish between the terms *endotherm* and *ectotherm.*

Answer on p. 112

Tested

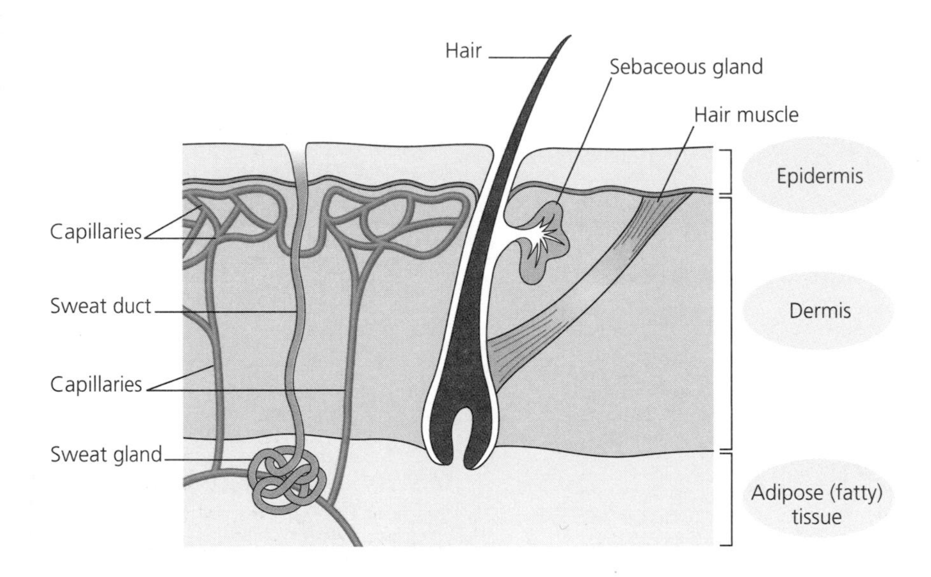

Figure 8.2 The basic structure of mammalian skin

If core temperature falls, this is also detected by the hypothalamus. The **heat gain centre** sends autonomic impulses that bring about:

- **vasoconstriction** (Figure 8.3) — skin arterioles constrict, redirecting blood along **shunt vessels** instead of flowing to the surface. In this way the heat in the blood is conserved
- **shivering** — rapid contraction of the skeletal muscles generates heat
- **pili-erection** — muscles attached to hair shafts contract, raising hairs and trapping an insulating layer of air against the skin. This is useless in humans, but effective in hairy mammals

In the short term, metabolic rate can be increased by the action of the hormone **adrenaline**, which increases blood glucose levels in preparation for action. It works by activating the enzymes that control **glycogenolysis** in much the same way as **glucagon**. In the longer term, the hormone

Glycogenolysis is the conversion of glycogen into glucose.

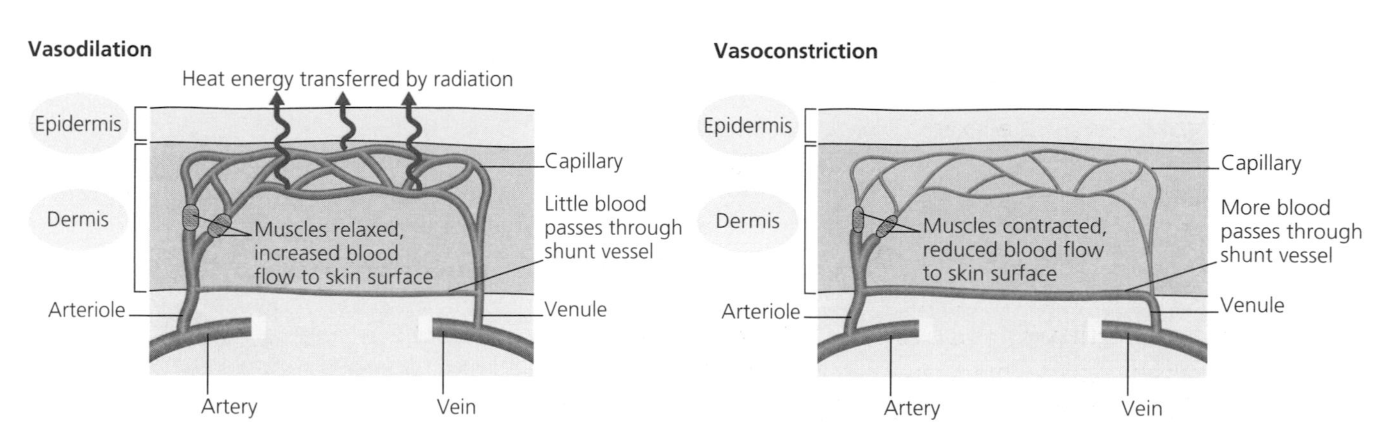

Figure 8.3 Vasodilation and vasoconstriction are brought about by arterioles, the only blood vessels that can significantly alter their size

thyroxine has the same effect. An increased metabolic rate increases heat generation.

If the temperature control mechanism breaks down, a positive feedback results (see page 81). If core temperature gets too high (over 40°C), enzyme activity increases, which controls respiration. An increased rate of respiration produces more heat, which makes temperature rise even further. It is not long before the lethal upper temperature is reached and the organism dies.

Control of blood glucose concentration

Processes involved

Revised

Glucose is absorbed into the bloodstream from the gut following carbohydrate digestion. After a meal there is usually more glucose than the body immediately requires, so the excess is stored as **glycogen** (Figure 8.4), with large amounts being stored in the **liver** and muscles.

Glycogen is a storage carbohydrate found in animals. It is a branched polymer of glucose that can be built up and broken down quickly according to the demands of the body.

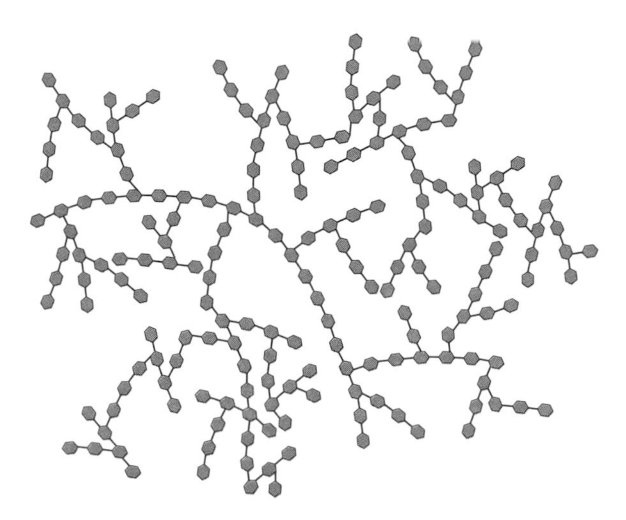

Figure 8.4 A glycogen molecule

Three key processes in glucose control are:

- **glycogenesis** — the synthesis of glycogen
- **glycogenolysis** — the breakdown of glycogen
- **gluconeogenesis** — when glycogen stores run low, new glucose can be generated by the conversion of amino acids and lactate

The role of insulin and glucagon

Revised

The **pancreas** (Figure 8.5) has a major role in the control of blood glucose. Small groups of cells, called **islets of Langerhans**, contain two types of hormone-producing cells:

- α-cells, which produce the hormone **glucagon**
- β-cells, which produce the hormone **insulin** (Figure 8.6)

The **islets of Langerhans** are small patches of endocrine (hormone producing) tissue in the pancreas.

These cells are unusual because they are both receptors and effectors. If blood glucose levels rise, it is detected by the β-cells which respond by secreting insulin into the blood. Insulin works by increasing the permeability of cell membranes to glucose. In this way, glucose can leave the blood and enter cells, thus lowering levels in the blood. If blood glucose levels fall, it is detected by the α-cells which respond by secreting glucagon into the blood. Glucagon works by activating the enzymes that break down glycogen, generating glucose that can pass into the blood and raise levels.

Examiner's tip

There are six G words in this topic, which can be confusing. Three are substances: glucose, glycogen and glucagon. Three are processes: glycogenesis, glycogenolysis and gluconeogenesis. To remember which is which, remember:
genesis = creation
lysis = splitting
neo = new

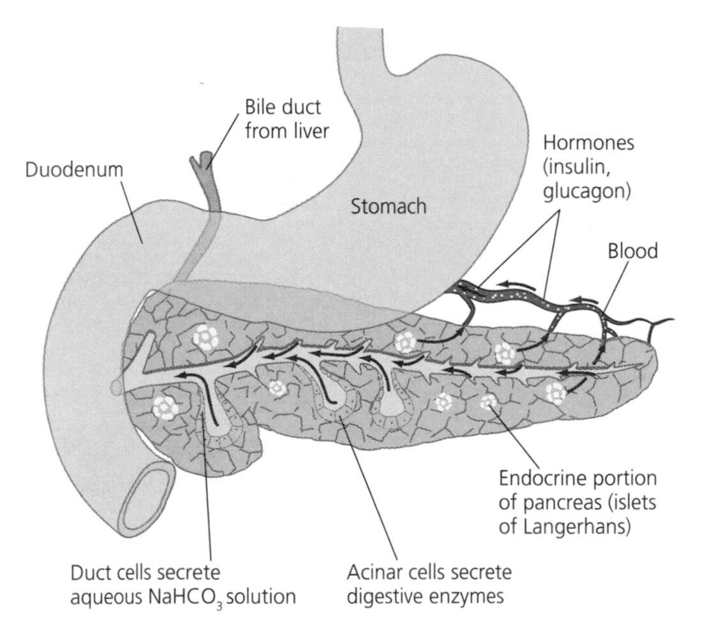

Figure 8.5 Most of the pancreas makes digestive juice, but the islets of Langerhans have a vital role in the control of blood glucose

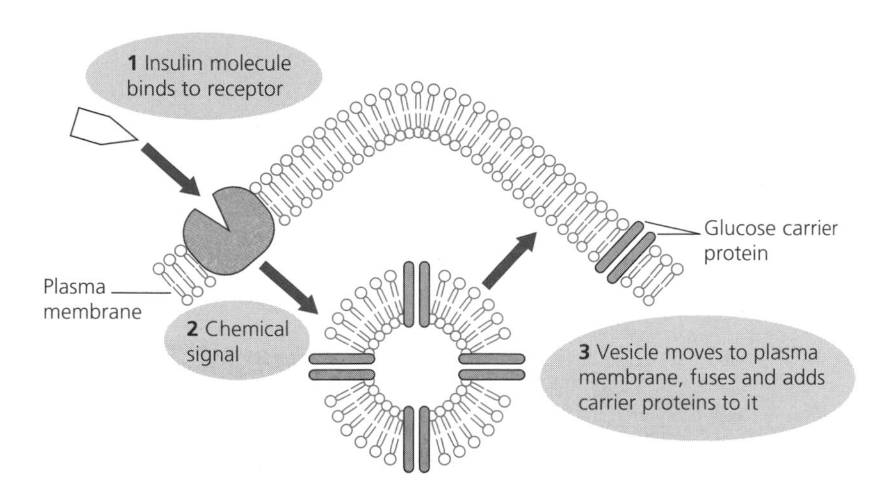

Figure 8.6 Insulin works by causing the cell to add extra glucose transport proteins to the cell surface membrane. Without insulin, the glucose transport proteins remain under the surface, where they cannot function

Hormones

Revised

There are two ways in which a hormone can work:

- Most hormones, including adrenaline and glucagon, work by the **second messenger model**. They do not enter the cells, but by binding to specific hormone receptors they alter the activity inside the cells.
- **Steroid hormones** are more direct. They go straight into the cell and change what happens in the nucleus, usually by activating specific genes.

The second messenger model

A **second messenger** is a substance found inside a cell that responds to the presence of a hormone outside the cell by activating a particular enzyme. The process is as follows:

1. The hormone is released by a gland and circulates in the blood to all parts of the body.
2. The hormone targets the cells with the correct receptor proteins in their cell surface membrane.
3. When the hormone combines with the receptor, an enzyme is activated in the cytoplasm.
4. The enzyme catalyses the conversion of **ATP** into **cyclic adenosine monophosphate (cAMP)**. The **first messenger** was the original hormone. As it cannot get into the cell, a second messenger, cAMP is needed to cause the desired effect inside the cell.
5. The cAMP activates certain enzymes. In the case of glucacon and adrenaline, it is the enzymes that catalyse the breakdown of glycogen.

This is an example of a **cascade reaction**, where a small amount of hormone can be amplified to bring about a significant effect. It is a chain of chemical events and each chemical can activate many more. As you saw in Unit 1, one enzyme can convert hundreds or thousands of substrate molecules into product in a very short time.

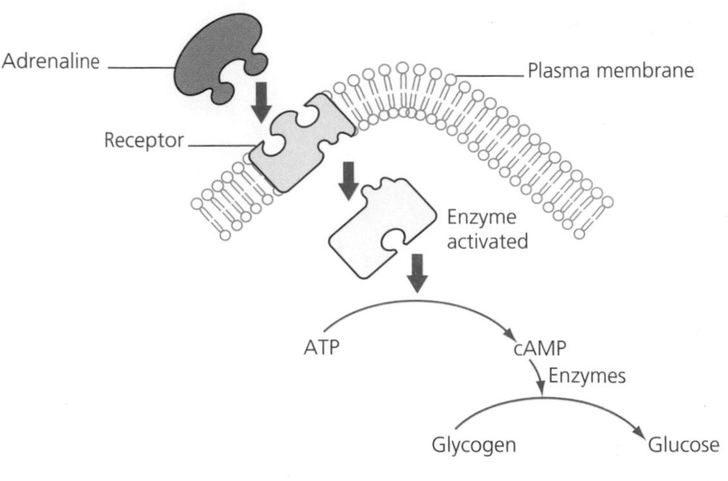

Figure 8.7 The second messenger model of adrenaline action. Glucagon works in much the same way, although the shape of the hormone and the receptor are different

Steroid hormones

Steroid hormones are lipids that are made from the cholesterol molecule. Being lipids, they can pass easily through cell membranes. Examples of steroid hormones are the male sex hormone testosterone and the female hormones oestrogen and progesterone (see page 87).

Steroid hormones pass directly into cells where they combine with a receptor in the cytoplasm. Once this active receptor complex has been made, it passes into the nucleus where it switches genes on or off (for more about the transcription of genes, see pages 93–94).

Diabetes

Revised

There are two types of **diabetes mellitus** (or 'sugar diabetes'): Type 1 and Type 2.

In **Type 1**, sufferers cannot make insulin and are usually treated by injections of the missing hormone. This is a lifelong condition that often starts in early childhood. The lack of insulin causes glucose to accumulate in the blood because it cannot pass into the cells quickly enough. The cells become starved of fuel and have to respire lipid as an alternative. Classic symptoms are:

- excessive thirst — glucose lowers the water potential of the blood
- excessive urination due to higher fluid intake
- glucose in the urine — the kidneys are unable to re-absorb all the glucose
- weight loss because the cells respire lipid reserves
- breath smells of ketones (a fruity smell) — a by-product of lipid metabolism

Type 1 diabetics have to monitor their glucose levels regularly. They get used to maintaining sensible blood glucose levels by injecting a mixture of slow- and fast-acting insulin and matching it to their sugar/carbohydrate intake.

In **Type 2** diabetes, sufferers make insulin but it may not be enough or the cells stop responding to it — a problem called **insulin resistance**. Type 2 is also called late-onset diabetes and it is associated with being overweight. It is usually treated by diet and exercise to control weight.

There are about 2 million diabetics in the UK, of which about 90% have Type 2.

Control of mammalian oestrous

What is the oestrous cycle?

Revised

The **mammalian oestrous cycle** is a sequence of events that involves the development of an egg or eggs within follicles and their release in the process of **ovulation**. The cycle is interrupted if pregnancy occurs.

There are four hormones that control the cycle (Figure 8.8):

- **follicle-stimulating hormone (FSH)**
- **luteinising hormone (LH)**
- **progesterone**
- **oestrogen**

Typical mistake

Some candidates think that all species have the same 28-day cycle as humans, but they don't. Many mammals synchronise their breeding cycle so that they ovulate at certain times of the year. Sheep, for example, come into season and ovulate in the autumn so that their lambs are born in spring.

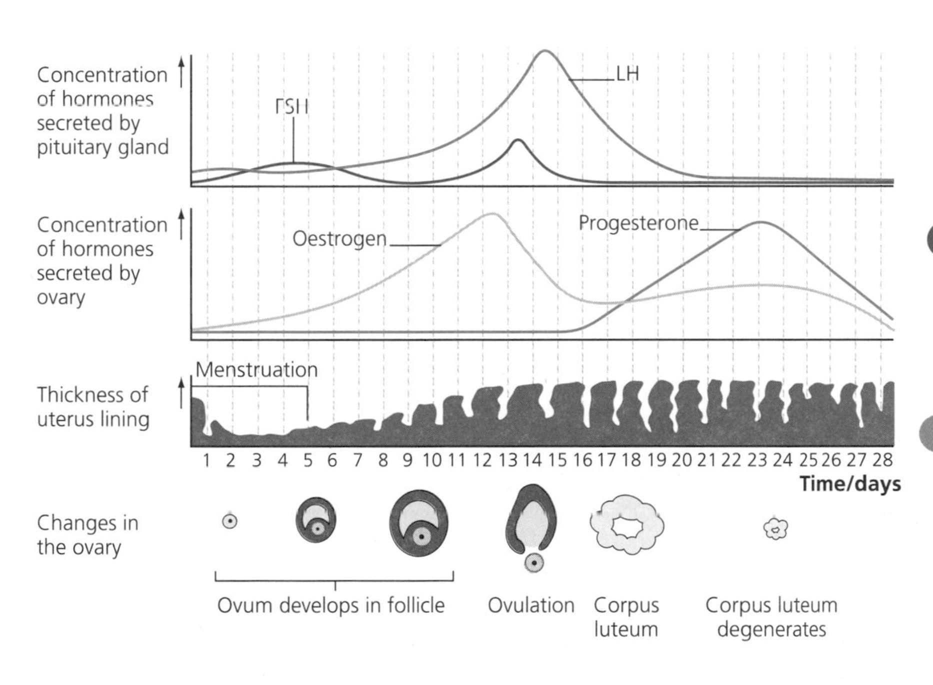

Figure 8.8 The events in the menstrual cycle are controlled by changing levels of four key hormones

Examiner's tip

You should be able to identify all of the four hormones by the shape of their curve on a graph.

Revision activity

Describe the curve of each hormone, so you can recognise which is which on an unlabelled graph. For example, 'spike in middle of cycle = LH'.

Events of the human oestrous cycle

Revised

Day 1

Menstruation starts. At the same time, FSH is secreted by the **pituitary gland**. This stimulates the development of an **ovarian follicle** in the **ovary**. A follicle is an ovum (egg cell) surrounded by layers of cells. The follicle begins to secrete oestrogen into the blood.

Day 5

Menstruation ends because the lining of the uterus (the **endometrium**) has been lost. Oestrogen stimulates the endometrium to regenerate.

Days 5–10

Oestrogen levels build up and the ovum matures in the follicle. Oestrogen inhibits LH production by the pituitary gland.

Days 10–14

As the ovum approaches maturity, oestrogen levels reach a threshold where the inhibition is reversed and a **positive feedback** begins. Oestrogen stimulates the production of LH, which in turn stimulates more oestrogen. The result is a surge in LH secretion — a spike on the graph — and a smaller surge of FSH. Within a few hours, ovulation takes place in which the ovum is released from a mature follicle. The remains of the follicle becomes the **corpus luteum**.

Days 15–21

The corpus luteum secretes progesterone as well as continuing to secrete some oestrogen. Progesterone stimulates the endometrium to mature and become ready for **implantation**. Progesterone inhibits FSH so that no new follicles develop while there is a corpus luteum present in the ovary.

If the egg is not fertilised, after about 28 days the corpus luteum shrinks and stops secreting progesterone. The inhibition of FSH is lifted and menstruation begins. A new follicle develops and the cycle starts again.

If the egg is fertilised, the early embryo implants in the uterus lining at about day 21, a week before the next period is due. The implanted embryo (a blastocyst) makes a hormone, **human chorionic gonadotrophin (HCG)**, which stimulates the corpus luteum to keep secreting progesterone. With high levels of progesterone, menstruation does not happen and no new follicles develop. The detection of HGC is the basis of pregnancy tests.

Now test yourself

Tested

3 Complete the following table.

Hormone	Made by	Target organ	Action
		Ovary	Causes a new follicle to develop
Oestrogen	Follicle	Various	Stimulates repair of the endometrium; combines with LH in a positive feedback to bring about ovulation
	Pituitary gland		
Progesterone		Various	Maintains the uterus lining; inhibits FSH production

4 The contraceptive pill works by maintaining high levels of oestrogen and progesterone. Explain how this prevents pregnancy occurring.

Answers on p. 112

Examiner's tip

The specification says that 'changes in the ovary and uterus lining are *not* required'. Questions on the oestrous cycle will focus on the mode of hormone action, positive and negative feedback and graphs that show the change in hormone levels.

Exam practice

1 The following graph shows temperature control in a fox and a lizard. Suggest which line, **A** or **B**, represents which organism. Explain your answer. [2]

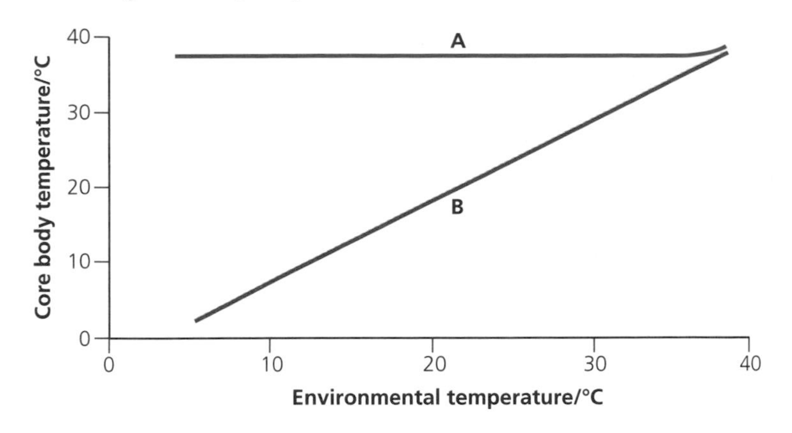

2 If an individual is suspected of being diabetic, a glucose tolerance test can be done. The patient has to fast for 12 hours (usually overnight) before being given a drink containing 75 g of glucose. The blood glucose levels are then monitored for 3 hours. The following graph shows the results for three individuals, **A**, **B** and **C**.

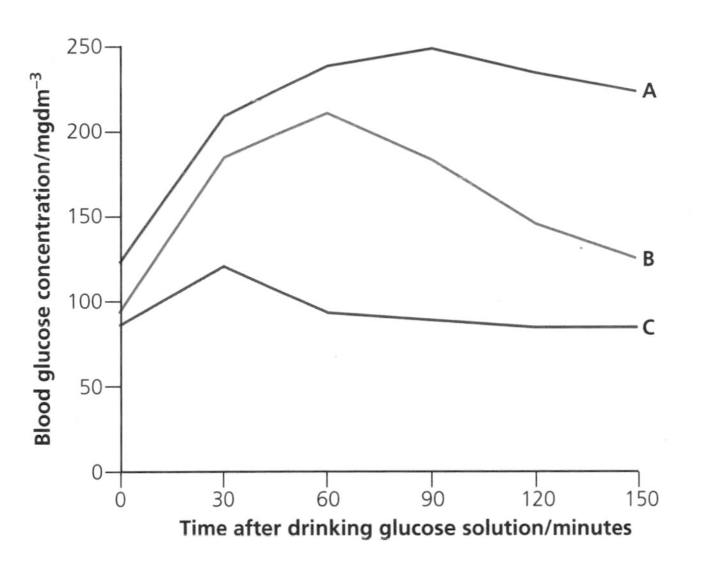

(a) Suggest why patients have to fast for 12 hours before the test. [1]

(b) Explain why there is a rise in all three graphs for the first 30 minutes. [1]

(c) Explain how a healthy individual responds to a rise in blood glucose. [3]

(d) Individual **A** is a Type 1 diabetic. Give three differences between the curves for individuals **A** and **C**. [3]

(e) Individual **B** is a borderline Type 2 diabetic. Suggest what treatment was advised. [1]

3 The following graph shows the levels of four hormones during one human menstrual cycle.

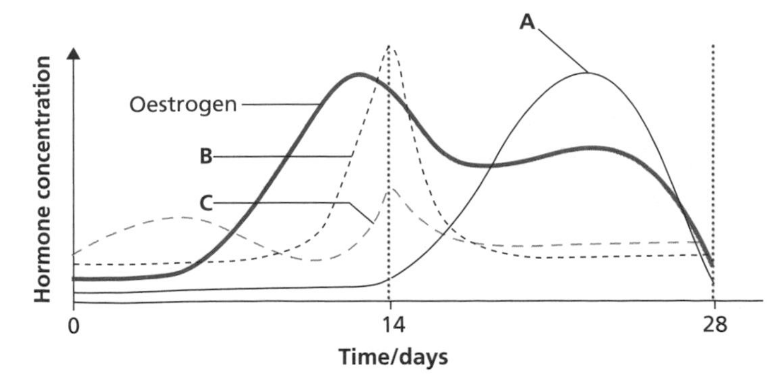

(a) Name the three hormones **A**, **B** and **C**. [3]

(b) Predict the time of ovulation. Explain your choice. [2]

(c) Explain how a positive feedback causes ovulation. [3]

(d) Pregnancy did not occur during this monthly cycle. Explain how the graph would be different if pregnancy had occurred. [1]

Answers and quick quizzes online

Online

Examiner's summary

By the end of this chapter you should be able to understand:

- ✔ Homeostasis involves control systems that maintain the internal environment within certain limits.
- ✔ Negative feedback is a 'detection-correction' mechanism that keeps systems steady.
- ✔ Positive feedback is a mechanism for change and can be damaging.
- ✔ Mammals are endotherms that can maintain a constant core body temperature.
- ✔ Reptiles are ectotherms whose body temperatures reflect those of the surroundings.
- ✔ The hypothalamus and the autonomic nervous system control the responses that maintain a constant body temperature.
- ✔ The factors that affect blood glucose levels, including the role of the liver.
- ✔ The role of insulin and glucagon in controlling blood glucose levels.
- ✔ The effect of adrenaline on glycogen breakdown and synthesis.
- ✔ The second messenger model of hormone action.
- ✔ The causes, symptoms and treatment of Type 1 and Type 2 diabetes.
- ✔ The roles of FSH, LH, progesterone and oestrogen in the mammalian oestrous cycle.
- ✔ The role of negative and positive feedback in the oestrous cycle.

9 Protein synthesis and its control

The genetic code

DNA and the genetic code

Revised

The **DNA** molecule was studied in Unit 2. It has two key functions: **replication** and **protein synthesis**. Replication was covered in Unit 2 and protein synthesis is covered here.

DNA codes for making proteins. The **base sequence** along the DNA molecules determines the **amino acid sequence** and therefore the structure of the protein (Figure 9.1).

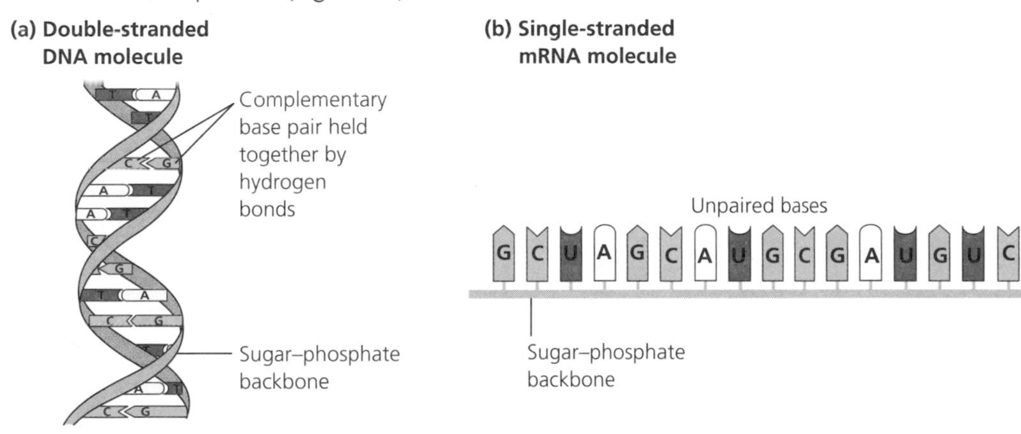

Figure 9.1 The key molecules in protein synthesis: (a) a section of DNA and (b) a section of mRNA. mRNA is essentially a mobile copy of a gene

Base triplets

Revised

The genetic code is a **triplet** code. Each amino acid is coded by a sequence of three bases. This is because if each of the four bases coded for an individual amino acid, it would only be possible to code for a total of four amino acids. If a group of two bases coded for an amino acid, there would be 4 × 4, or 16 possible combinations. As 20 amino acids are found in proteins, this still would not be enough. A group of three bases allows for 4 × 4 × 4, or 64 combinations, which is more than enough to code for all 20 amino acids.

As in DNA replication, the key to protein synthesis is **complementary bases**. When the two strands of DNA are separated, the sequence on the **gene** can be copied by adding complementary RNA nucleotides. These are similar to DNA nucleotides but the base thymine (T) is replaced by uracil (U). Therefore, if a section of the gene reads TAT GCG TTA, the complementary RNA sequence is AUA CGC AAU.

A **gene** is a length of DNA that codes for making one polypeptide or protein. A gene always codes for making a polypeptide, but some proteins consist of more than one polypeptide, in which case it will be coded for by more than one gene.

Universal, non-overlapping and degenerate

Revised

The genetic code is:

- **universal** — the same **codons** code for the same amino acids in all known organisms
- **non-overlapping** — a sequence of CCTGGC is just two codons, CCT and GGC. If the code overlapped there would be codons of CCT, CTG, TGG and GGC. Each base is used once only
- **degenerate** — there are 64 different codons but only 20 amino acids, so there are spare codons. Most amino acids have more than one codon and some, such as leucine, have as many as six (Figure 9.2)

A **codon** is a sequence of three bases on an mRNA molecule that codes for an amino acid, although there are three codons that do not code for any amino acid. They act as 'stop' signals and indicate the end of a particular protein.

Examiner's tip

You don't have to remember any codon or amino acid combinations. These will always be provided in the exam question.

First position	**Second position**				**Third position**
	T	C	A	G	
T	Phenylalanine Phenylalanine Leucine Leucine	Serine Serine Serine Serine	Tyrosine Tyrosine (stop) (stop)	Cysteine Cysteine (stop) Tryptophan	T C A G
C	Leucine Leucine Leucine Leucine	Proline Proline Proline Proline	Histidine Histidine Glutamine Glutamine	Arginine Arginine Arginine Arginine	T C A G
A	Isoleucine Isoleucine Isoleucine Methionine	Threonine Threonine Threonine Threonine	Asparagine Asparagine Lysine Lysine	Serine Serine Arginine Arginine	T C A G
G	Valine Valine Valine Valine	Alanine Alanine Alanine Alanine	Aspartic acid Aspartic acid Glutamic acid Glutamic acid	Glycine Glycine Glycine Glycine	T C A G

Figure 9.2 The genetic code using DNA bases

Now test yourself

Tested

1 How can you tell that Figure 9.2 contains DNA codes and not RNA codes?

2 Use Figure 9.2 to complete the following table.

DNA sequence	AAT		GTC
mRNA sequence		GUA	
Amino acid			

Answers on p. 112

Structure of mRNA and tRNA

Revised

There are two types of RNA in protein synthesis:

- **messenger RNA (mRNA)** is a long, single-stranded polynucleotide chain that is assembled on a gene
- **transfer RNA (tRNA)** is a small, clover-leaf shaped 'fetch and carry' molecule that brings amino acids to the site of protein synthesis

Transfer RNA (tRNA) is a type of nucleic acid that is important in assembling amino acids in the correct order during protein synthesis.

Table 9.1 Comparing DNA, mRNA and tRNA

Feature	DNA	mRNA	tRNA
Sugar	Deoxyribose	Ribose	Ribose
Bases	A, C, G, T	A, C, G, U	A, C, G, U
Number of strands	Two	One	One
Hydrogen bonds?	Yes	No	Yes
Number of nucleotides	Millions	Hundreds or thousands — it depends on the size of the gene	About 75

Now test yourself Tested

3 Give three differences between the structures of mRNA and tRNA.

4 How many different types of tRNA exist in cells? Explain your answer.

Answers on p. 112

Polypeptide synthesis

Transcription

Revised

Transcription is the first stage of protein synthesis. DNA cannot leave the nucleus, but proteins are built on the **ribosomes**. As a consequence, the genetic code must be copied in the nucleus and transferred to the ribosomes. Transcription is the process of copying the genetic code by making mRNA from DNA (Figure 9.3).

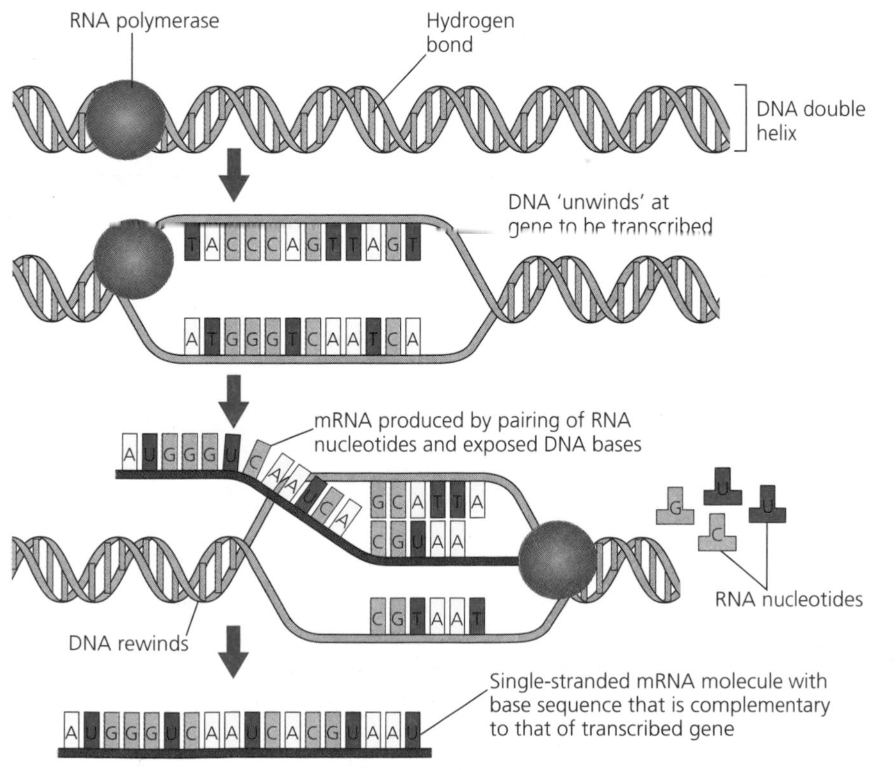

Figure 9.3 Transcription

1 The enzyme **RNA polymerase** attaches to the start of the gene.

2 The DNA 'unwinds' as the hydrogen bonds are broken.

3 The RNA polymerase moves along the gene, catalysing the addition of **complementary nucleotides**.
4 When complete, the mRNA passes out of the nucleus, through the nuclear pores and into the cytoplasm where it attaches to a ribosome.

Typical mistake

Many candidates state that RNA polymerase adds complementary bases, whereas the term should be complementary nucleotides. It is the bases that actually join, but they also have a sugar and a phosphate attached.

Introns and exons

A gene in a **eukaryotic cell** consists of non-coding sequences (**introns**) and coding sequences (**exons**). During transcription, the entire base sequence of a gene is transcribed to produce **pre-mRNA**. This includes both the introns and the exons. Before it leaves the nucleus, the pre-mRNA is edited and the introns are removed. The exons are then spliced together to produce mRNA that carries only the coding sections of the gene.

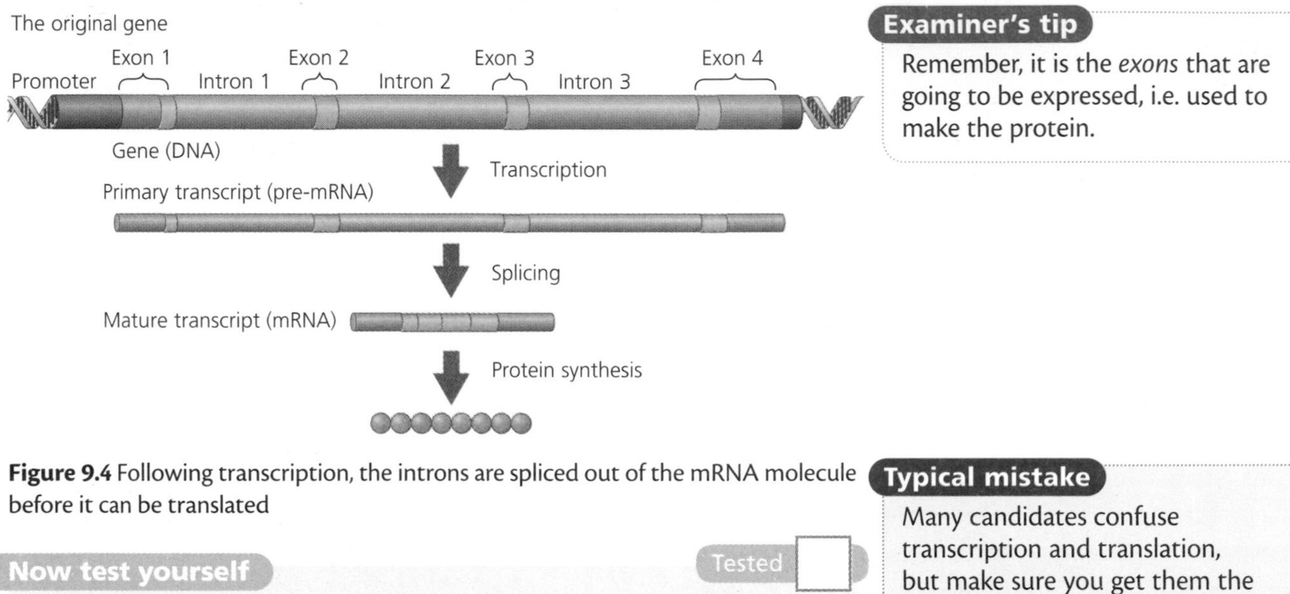

Figure 9.4 Following transcription, the introns are spliced out of the mRNA molecule before it can be translated

Examiner's tip

Remember, it is the *exons* that are going to be expressed, i.e. used to make the protein.

Now test yourself Tested

5 What is the difference between pre-mRNA and mature mRNA?

Answer on p. 112

Typical mistake

Many candidates confuse transcription and translation, but make sure you get them the right way round. Remember that *-cription* comes before *-lation* both alphabetically and in biology.

Translation

Revised

Translation is the second stage of protein synthesis. It involves assembling a protein by joining amino acids together according to the sequence encoded on the mRNA. The key organelle is the ribosome, which can be thought of as a giant enzyme that holds all the different components together so that the process can happen. The tRNA molecules are relatively small, with two key features (Figure 9.5):

- an **anticodon** consisting of three bases
- an **amino acid binding site**

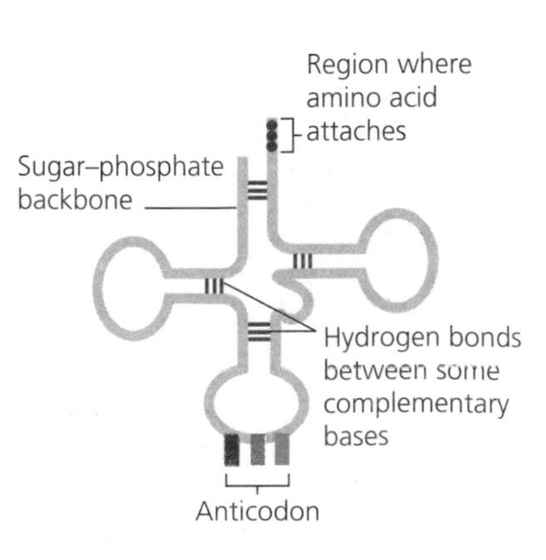

Figure 9.5 A tRNA molecule

The anticodon and the amino acid are always matched. For example, the mRNA codon AUG codes for the amino acid methionine. When this codon is translated, a tRNA molecule with the anticodon UAC arrives carrying a methionine at the other end. The vital steps are:

1 Ribosomes have two codon-binding sites. The first two codons on the mRNA molecule attach to the binding sites.
2 The first codon is translated. It reads AUG. A tRNA molecule with the anticodon UAC arrives carrying a methionine at the other end. The amino acid is held in place (Figure 9.6).

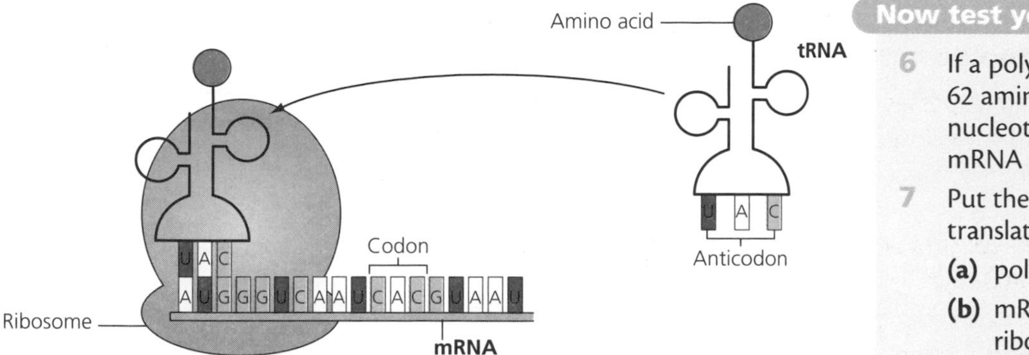

Figure 9.6 Translation, steps 1–2

3 The second codon is translated. The second amino acid is brought in by the tRNA molecule and held alongside the first one.
4 An ATP molecule attaches and is hydrolysed. The energy released is used to form the **peptide bond** between the two amino acids.
5 The mRNA moves alone the ribosome, one codon at a time. The **polypeptide** grows as each codon is translated (Figure 9.7).

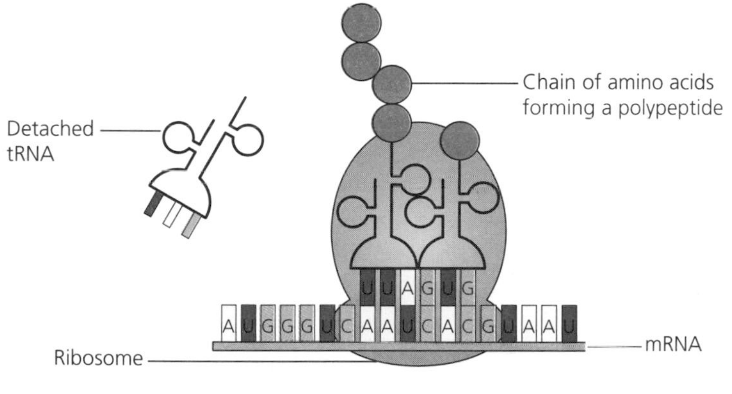

Figure 9.7 Translation, steps 3–5

Now test yourself

6 If a polypeptide consists of 62 amino acids, how many nucleotides will the mature mRNA have?
7 Put the following events of translation in order.
(a) polypeptide grows
(b) mRNA attaches to ribosome
(c) peptide bonds form
(d) ATP is split
(e) tRNA delivers amino acids and holds them alongside each other
(f) the first two codons are translated together
(g) mRNA moves along ribosome

Answers on p. 112

Tested

Revision activities

Draw your own cartoon or storyboard to explain translation — making the components from card works well. You will need a large ribosome, one mRNA with bases, at least two tRNAs with anticodons and an equal number of amino acids.

Gene mutation

Which gene mutations affect the organism?

Revised

Gene mutations occur because of mistakes in **DNA replication** that result in a changed base sequence. This changes the **genotype** of the organism and may be inherited. Many mutations do not affect the organism because:

- some take place in the non-coding DNA between genes
- some take place in the introns (non-coding sequences) within genes
- some will still code for the same amino acid — the genetic code is degenerate. For example, if the codon GUU mutates to GUC, GUA or GUG, it will still code for the amino acid valine
- some will cause a change in the amino acid sequence, but this will not significantly change the tertiary structure. The protein is still the right shape and will still function in the organism

Therefore, the only gene mutations that affect organisms are those that bring about significant changes in the structure of the protein. There are two main ways in which a base sequence can be altered:

- **substitution**, in which one base is substituted for another. This is also called a **point mutation**. Only one codon is changed, but this can still have a significant effect on the protein
- **deletion**, in which one base is lost and there is a **frame shift** — all the bases move along in one direction and therefore many codons are changed

In Figure 9.8, the top two rows show the original sequence. The second two rows show the effect of a substitution: only one codon and one amino acid are changed. The third two rows show the effect of deleting the red letter A: a frame shift results so that all codons and all amino acids are changed.

Original base sequence on mRNA	AGA	UAC	GCA	CAC	AUG	CGC
Encoded sequence of amino acids	Arginine	Tyrosine	Alanine	Histidine	Methionine	Arginine
mRNA base sequence after base substitution	AGU	UAC	GCA	CAC	AUG	CGC
Encoded sequence of amino acids	Serine	Tyrosine	Alanine	Histidine	Methionine	Arginine
mRNA base sequence after base deletion	AGU	ACG	CAC	ACA	UGC	GCx
Encoded sequence of amino acids	Serine	Threonine	Histidine	Threonine	Cysteine	Alanine

Figure 9.8 The effects of mutation

Mutagenic agents

Revised

Gene mutations occur randomly, at a slow and steady rate and in any part of the DNA. Most of these mistakes are spotted and corrected by a 'proofreading' mechanism within the cell. However, the rate of mutation can be increased by **mutagenic agents**, in which case the proofreading may not spot them all. Mutagenic agents include:

- some chemicals including benzene, mustard gas and bromine/bromine compounds
- ionising radiation (gamma and X-rays)
- ultraviolet light
- biological agents such as some viruses and bacteria

Cell division and tumours

Revised

The control of **cell division** is vital. Cells should only divide when needed, so that body tissues replace themselves at the correct rate. If a cell's control system breaks down and cell division (mitosis) happens too fast, a **tumour** can result (Figure 9.9).

The rate of cell division is controlled by genes called **proto-oncogenes**. If these genes mutate into **oncogenes**, cells may start to divide too quickly. There is a back-up system in the form of **tumour suppressor genes**, which prevent rapid cell division or cause the death of the cell if the damage cannot be repaired. If cell division occurs at the centre of the tumour and does not spread, it is known as **benign**. If dividing cells are at the edge of the tumour and likely to break off and set up secondary tumours, it is **malignant** — that is cancer.

We accumulate these mutations throughout our lives and if the proto-oncogenes mutate and the tumour suppressor genes mutate, a tumour can result. There is another back-up system: the immune system.

White cells can detect abnormal cells in the same way that they detect pathogens so that small tumours are destroyed.

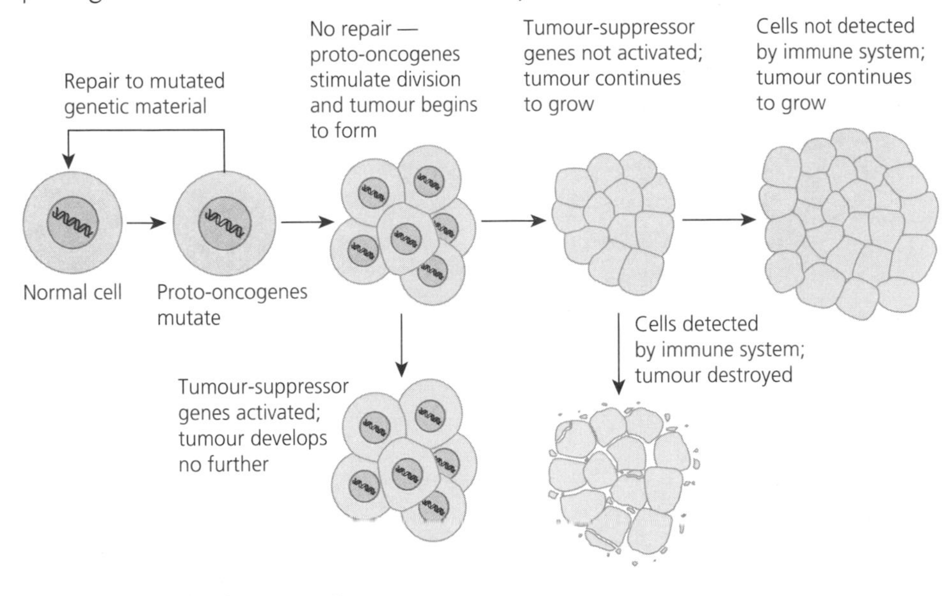

Figure 9.9 The development of tumours

Gene expression

How cells specialise

Revised

The key to cell specialisation is the **selective activation of genes**. In muscle cells, for example, the genes that code for the contractile proteins actin and myosin need to be activated. There is a huge amount of DNA in every cell and the vast majority of it is never translated into proteins. In any particular cell, only specific genes are active.

Stem cells are unspecialised cells with two key properties:

- **self-renewal** — the ability to divide again and again
- **potency** — the ability to become specialised. Usually, this is a one-way process; as the cells specialise, they lose their potency. This requires the stem cells to be **totipotent**

Totipotent cells occur only for a limited time in mammalian embryos. In an adult there are some tissues, notably brain and bone marrow, that contain **multipotent** stem cells. These have the potential to develop into a limited number of cell types, so they are not as flexible as totipotent cells.

Totipotent cells have the potential to develop into all cell types that make up an organism.

Multipotent cells can differentiate into a number of cells, but only those of a closely related family of cells.

There is great potential for using stem cells in medicine. For example:

- replacement β-cells to treat Type 1 diabetics
- brain cells for Parkinson's disease
- new skin cells for burns victims

However, embryonic stem cell research is controversial. There is no shortage of human embryos because infertility treatment creates more than is needed. Some people say that the embryo is a human life with the potential to grow into a new individual and it therefore should be given rights. Others argue that an embryo is a tiny ball of cells without a name or a nervous system that would otherwise simply be destroyed.

Examiner's tip

The specification says that you should be able to 'evaluate the use of stem cells in treating human disorders'. Evaluate means to look at both sides.

Stem cells in plants

Revised

In mammals, only early embryonic stem cells are totipotent, but most plant cells remain totipotent in adulthood. This allows whole plants to be **cloned** from a simple tissue sample or cutting. For example, a piece of cabbage leaf, if placed in sterile conditions, can develop into a whole new cabbage plant in a short period of time. This is useful because:

- plants with desirable characteristics can be reliably cloned
- some valuable crops — orchids, for example — can be grown quickly with a minimum of space and time

Cloning is an example of asexual reproduction. The advantage is that there is no mixing of genetic information and so the genotype is always preserved unless there is a mutation.

Control of gene expression

Revised

Gene expression involves the following flow of genetic information: DNA → mRNA → polypeptide. It is possible to control the expression of genes at any stage in the process, such as:

- preventing transcription
- interfering with the splicing of the introns
- destroying the mRNA
- interfering with translation (e.g. by blocking the ribosome)
- preventing a polypeptide from being turned into a functional protein

Regulation of transcription and translation

Transcription of target genes

Revised

Transcription of a gene starts when the RNA polymerase enzyme binds to the start of the gene and begins to make mRNA. However, this can only happen when all the **specific transcriptional factors** are in place. These factors combine to form the **transcription initiation complex (TIC)**. Some the factors are present in the **cytoplasm**, moving into the **nucleus** when needed, and some of them come from outside the cell. Some hormones act as transcriptional factors.

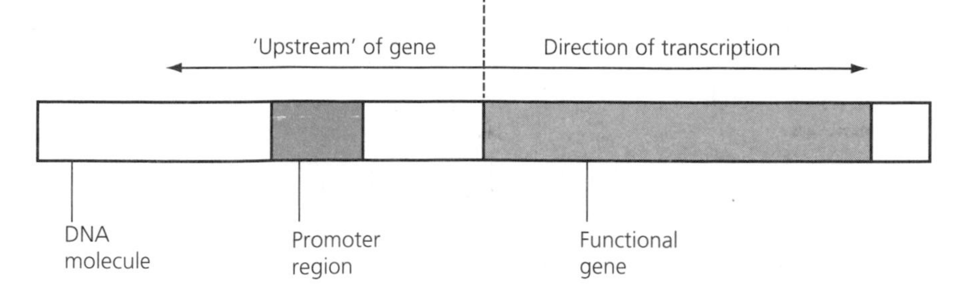

Figure 9.10 In eukaryotes, all genes have a promoter region that is described as being upstream of the gene

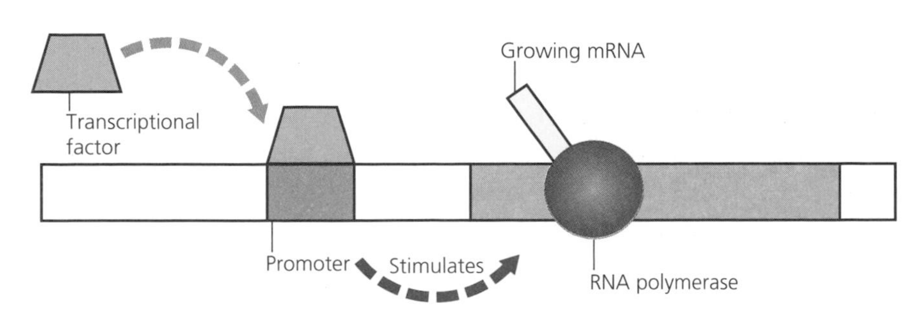

Figure 9.11 The functional gene contains the code for making the polypeptide, but it will not be transcribed unless the correct transcriptional factors attach to the promoter

Oestrogen and gene transcription

Revised

Oestrogen is a steroid hormone (see page 86). As steroids are lipids, they pass through the cell membrane. Oestrogen passes into the cytoplasm of target cells, where it combines with an oestrogen receptor called ERα. This forms an active ERα complex that passes into the nucleus and acts as a transcriptional factor for many different genes.

Small interfering RNA

Revised

Small interfering RNA (siRNA) is a short, double-stranded molecule of RNA that interferes with the expression of a specific gene by breaking down the mRNA and thereby preventing translation.

One molecule of siRNA combines with several proteins to form an RNA–protein complex. One of the strands of the siRNA comes away and the other is used by the RNA–protein complex to seek out and bind to the mRNA that needs silencing. It is another example of complementary base pairing. The mRNA is cut by the proteins and therefore cannot be translated.

It is thought that siRNA evolved as a defence against viral attack and it has great potential in medicine because it can prevent the expression of harmful genes. It is one example of the molecules used in the process of RNAi (RNA interference).

Examiner's tip

It is easy to go into too much detail about the way siRNA works. Exam questions will not ask for details beyond the basic mechanism and its potential. Most questions will use siRNA to test your understanding of protein synthesis or complementary base pairing.

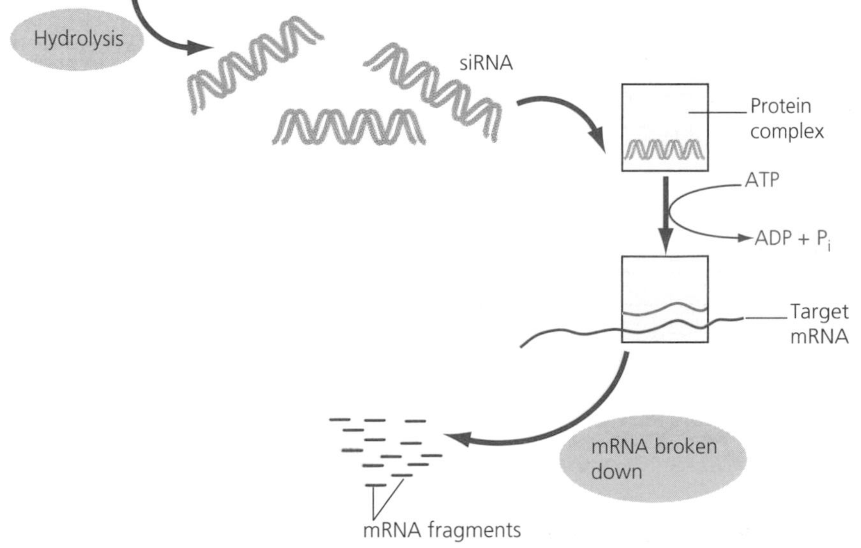

Figure 9.12 Small interfering RNA (siRNA) are short, double-stranded molecules of RNA

Exam practice

1 The following base sequence is taken from within a gene: AAGGCTCCATTG.

(a) What is the maximum number of amino acids that could be coded for by this sequence? [1]

(b) What is the minimum number of amino acids that could be coded for by this sequence? Explain your answer. [2]

(c) Write down the corresponding mRNA sequence that would result from transcription. [1]

(d) In a mutation, the red letter A was substituted.

(i) How many triplets would be changed? [1]

(ii) Give two reasons why this mutation might not affect the organism. [2]

(e) The whole gene was analysed and the percentage of bases is shown in the following table. Fill in the missing value. [1]

Base	A	T	G	C
Percentage in gene	20	29	22	

2 Explain how oestrogen affects gene expression in a target cell. [5]

Answers and quick quizzes online

Online

Examiner's summary

By the end of this chapter you should be able to understand:

- ✔ The structure and function of mRNA and tRNA.
- ✔ The genetic code consists of base triplets which code for specific amino acids.
- ✔ The genetic code is universal, non-overlapping and degenerate.
- ✔ The events of transcription as the production of mRNA from DNA and the removal of introns.
- ✔ The events of translation.
- ✔ Gene mutations are changes in the base sequence that result from faults in DNA replication.
- ✔ The role of mutations in the development of tumours.
- ✔ The difference between totipotent and multipotent stem cells.
- ✔ The potential uses and ethical issues relating to stem cells.
- ✔ The potential of stem cells in plants.
- ✔ The effect of oestrogen on gene transcription.
- ✔ The role of siRNA in blocking the expression of a specific gene.

10 Gene technology

Gene cloning and transfer

Key concepts

Revised

Gene technology, or **genetic engineering**, involves the manipulation of the DNA molecule. The key processes include:

1. finding genes
2. cutting them out
3. finding out their base sequence
4. cloning them
5. putting them into different cells and organisms so that they are expressed

The human genome

Revised

The ultimate goal is to understand the human **genome** and that of many other species. The aim is to find out where all the genes are, what they do and how they interact to build and maintain an organism. This a fast-moving area of science and the technology advances almost on a daily basis. The principles covered in this area of the specification are all sound, but technology moves on. Many of the processes that used to be labour-intensive are now automated and rapid. There are machines that can sequence an entire human genome — 3 billion base pairs — in a couple of hours.

A **genome** is all the genetic material in a single cell from an organism. In humans, we know it consists of 23 chromosomes, over 3 billion base pairs and our current best estimate is that there are about 23 000 genes.

Examiner's tip

The exam may ask you about the basics of genetic engineering, but you will not be expected to know about the very latest developments.

Table 10.1 The genetic engineering toolkit

Tool	Job
Restriction endonuclease	An enzyme that cuts DNA at specific recognition sites; usually produces sticky ends rather than clean cuts
Ligase	An enzyme that joins two pieces of DNA, such as complementary sticky ends to form recombinant DNA
Reverse transcriptase	An enzyme that makes a DNA molecule from the corresponding mRNA (i.e. transcription in reverse); found in retroviruses such as HIV/AIDS; DNA made this way is called complementary DNA (cDNA)
DNA or gene probes	Radioactive or fluorescently labelled fragments of DNA that seek out and bind to a target sequence; used to detect the presence of particular genes, for example in medical diagnosis
Plamids	Small, circular pieces of DNA found in bacteria; used as vectors to put genes into bacteria

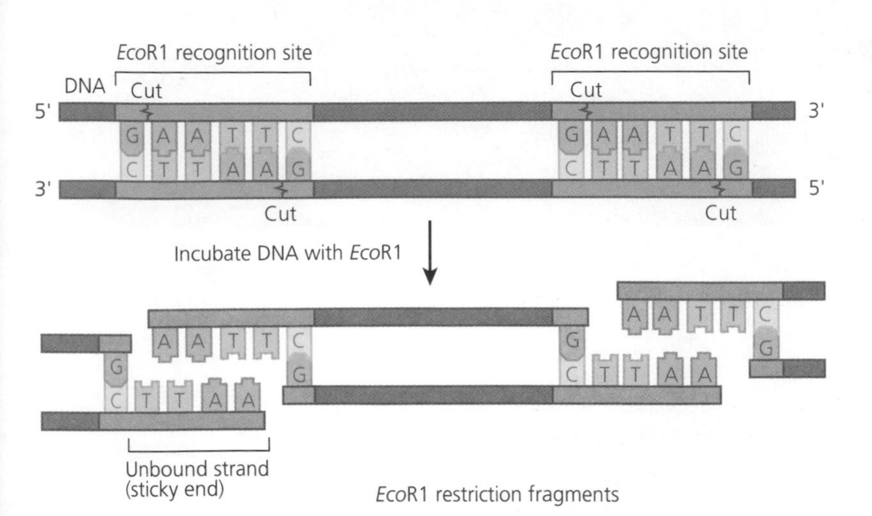

Figure 10.1 The restriction enzyme *Eco*R1 cuts DNA at the specific recognition sequence GAATTC. This leaves two sticky ends that can be joined to any other piece of DNA that has been cut by the same enzyme

In English, a palindrome is a word or phrase that is spelt the same forwards as backwards, such as *gnu dung*. It is not the same in biology. A **palindromic** DNA base sequence is the same on the other strand, reading in the other direction. For example, the sequence GCTAGC would read CGATCG on the other strand, which is the identical sequence backwards.

In vivo and *in vitro* techniques

Revised

In vivo means 'in life' and refers to experiments carried out on living materials rather than in test-tube conditions. ***In vitro*** means 'in glass' and refers to experiments carried out in test-tube conditions or a petri dish rather than on living materials.

In vivo cloning involves putting DNA into a living cell so that the gene is copied each time the cell divides. In addition, the gene can be expressed because the whole cell contains ribosomes and all the other components needed to make proteins. *In vitro* cloning involves making copies of DNA using the **polymerse chain reaction (PCR)** (see page 103).

The **polymerase chain reaction (PCR)** is a process used by biologists to make large amounts of identical DNA from very small samples.

Table 10.2 Advantages and disadvantages of *in vivo* and *in vitro* gene cloning

	Advantages	Disadvantages
In vivo cloning (in whole cells)	More accurate — fewer mistakes because the cell has 'proofreading' correction mechanisms Can copy unknown DNA Can reliably copy large fragments of DNA Can express the clones gene and make the protein	Takes time Requires a large sample Requires more purification — DNA needs to be extracted from the cell
In vitro cloning (PCR)	Quick — one cycle takes about 3 minutes Works on minute quantities Works on partially decomposed DNA — from old remains, perhaps Simple purification from solution	Lots of copying mistakes — no 'proofreading' correction mechanisms Not reliable on large fragments Does not work on unknown DNA because complementary primers are needed Cannot make the protein encoded for by the gene

Where do we get DNA fragments from?

Revised

DNA molecules are huge, so genetic engineering usually involves working with fragments of DNA. There are three ways to make DNA fragments:

- work backwards from mRNA, using reverse transcriptase to make **complementary DNA (cDNA)**
- cut out specific pieces of DNA, such as genes, using restriction enzymes
- clone existing pieces of DNA using the PCR

Restriction enzymes come from bacteria where they are weapons against viral infection. Their full name, **restriction endonuclease**, refers to the fact that they *restrict* viral growth by cutting *within* the *nucleic acid*.

Polymerase chain reaction (PCR)

Revised

The polymerase chain reaction is gene cloning in a test tube. It is an effective and quick way of amplifying small samples of DNA so there are multiple copies of useful genes or enough DNA for a DNA profile, for example. You need the following:

- the DNA template to be copied
- DNA polymerase to copy the DNA; a thermostable enzyme is needed
- a supply of nucleotides
- primers — short sections of single-stranded DNA that allow the enzyme to attach and start copying

The thermostable enzyme commonly used is Taq polymerase. It is extracted from *Thermus aquaticus*, a bacterium found in hot volcanic springs. The enzyme is not denatured even at 94°C, so it can function at all the different temperatures encountered in the PCR cycle. The key steps in the PCR are:

1. **Denaturation** — heat the mixture to 94°C to denature the DNA; the hydrogen bonds break and the two strands separate, yielding single-stranded DNA molecules.
2. **Annealing** — cool the mixture to 50–60°C to anneal (attach) the primers to the single-stranded DNA template.
3. **Extension** — heat the mixture to 74°C; the Taq polymerase moves along the DNA, catalysing the addition of complementary nucleotides.

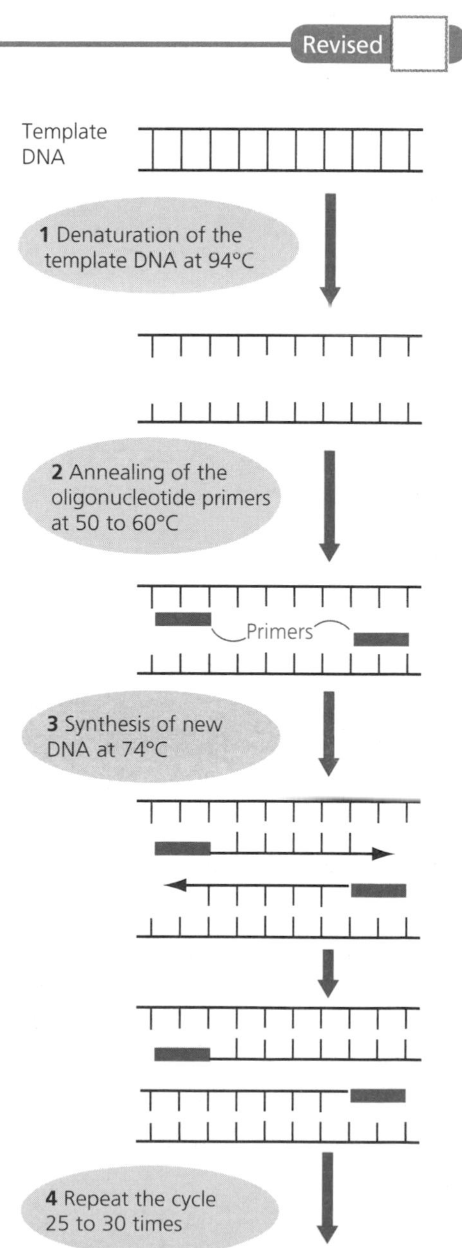

Figure 10.2 The key steps in the PCR. The whole process is done in a machine called a thermal cycler, which can go through the cycle in 3 minutes. 25 cycles takes just over an hour, by which time a million copies will have been made

Now test yourself

Tested

1 Starting with a single piece of DNA, how many pieces would you have after six PCR cycles?

Answer on p. 112

Recombinant DNA technology

Revised

Recombinant DNA is the technology that transfers DNA, putting genes from one species into another. Examples include inserting:

- the human insulin gene into bacteria so they make human insulin for diabetics
- a gene that makes vitamin A into rice crops so you get vitamin A-rich Golden Rice
- herbicide resistance genes into crops so that spraying kills the weeds but not the crops

Recombinant DNA results from joining the DNA of two different species, e.g. bacteria and human.

Using the insulin example, the main steps are:

1. Find the target gene and cut it out using a restriction endonuclease.
2. Clone it — make lots of copies using the PCR.
3. Put the gene into a **vector**. This involves getting a **plasmid** and cutting it with the same restriction endonuclease that was used to cut out the gene. The complementary **sticky ends** on the plasmid and the gene are joined with the enzyme **ligase**.
4. Mix the **recombinant plasmids** with bacteria and treat the mixture so that some of the bacteria take up the plasmid.
5. Find the bacteria that have adopted the new plasmid.
6. Culture the **transgenic** bacteria in a sterile environment (grow them in a large vat with a supply of nutrients and oxygen) so that they multiply and express the gene, producing insulin which can then be extracted and purified.

Sticky ends are unpaired base sequences at the ends of a piece of DNA.

Transgenic is a term that describes an organism that contains DNA transferred from another organism.

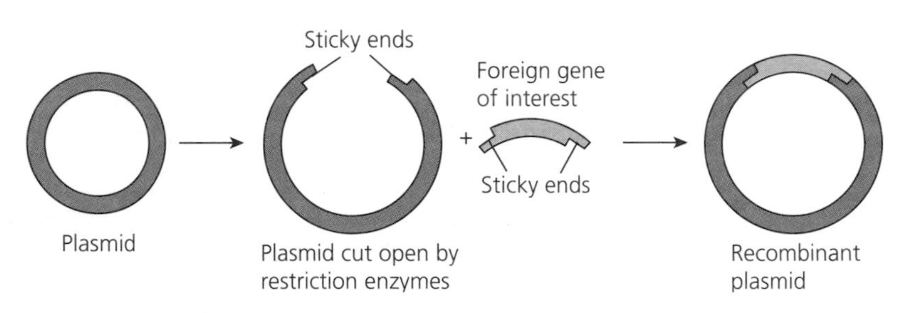

Figure 10.3 Plasmids are useful vectors for 'smuggling' foreign DNA into bacteria

2 What is the importance of sticky ends?

Answer on p. 112

Tested

Finding the recombinant bacteria

This process is unreliable — most bacteria do not take up the new DNA — so finding the recombinant ones is important. A common method is to add genes for antibiotic resistance so that only the bacteria that have accepted the gene of interest will survive when grown on a medium that contains the relevant antibiotic.

Possible objections to gene cloning and transfer

Revised

There are several objections to DNA technology:

- **Genetically modified (GM)** crops contain genes that can be transferred to other organisms. This may give rise to pesticide-resistant super-weeds that are difficult to control and may upset the ecosystem.
- Cloning whole organisms is unreliable and there are many stillbirths and animals born with defects. It took 200–300 attempts to get Dolly the sheep and when dealing with a sophisticated animal with a nervous system that amount of suffering is unacceptable.

Gene therapy

The use of gene therapy

Revised

Genetic diseases cannot be cured. People with **sickle cell anaemia** or haemophilia, for example, have the defective alleles in every cell in their body and that cannot be changed.

However, **gene therapy** brings the promise of treatment by using multiple copies of healthy, working alleles to supplement defective ones. In cystic fibrosis, for example, it may be possible to introduce multiple copies of the working CFTR allele into the lungs. The patient inhales an aerosol — similar to an asthmatic with an inhaler — containing the healthy gene in a vector such as a liposome or virus. The gene is absorbed into the epithelial cells of the lungs where it is expressed and used to make the missing protein.

This approach is still at the trial stage and few success stories have been reported. One of the problems is how to deliver the alleles to the affected cells, most of which are more difficult to get at than the lining of the lungs. However, gene therapy seems to have great potential for the future.

Examiner's tip

When asked to state an objection to a particular treatment such as gene therapy, your answer needs to include some biology. 'We should not play God' is never a good answer. Problems with gene therapy include the possibility that the virus will cause disease or that the long-term effects of altering a cell's genome are difficult to predict: tumours could result, for example.

Now test yourself

3 Suggest why this aerosol treatment would need to be repeated regularly.

Answer on p. 112

Tested

Medical diagnosis

Predisposition to diseases

Revised

There is an increasing demand for tests that can tell whether individuals possess certain alleles. In addition to well-known genetic diseases such as cystic fibrosis, we are starting to identify many new alleles that make people **predisposed** to diseases such as certain types of cancer.

Predisposed is a term frequently used in medicine. It means that the possession of certain genes makes an individual more *likely* to develop a particular disease.

DNA probes

Revised

DNA probes can find disease-causing alleles. The probe is a fragment of DNA that is complementary to part of the base sequence of the gene in question. The probe is added to a tissue or cell sample and binds to (or **hybridises with**) the complementary sequence. The fragment is **labelled** with a radioactive or fluorescent marker so that the gene shows up. If the **screening test** is positive, knowledge that an individual possesses a particular gene gives them the opportunity to:

- make certain lifestyle changes — if they find they are predisposed to high blood cholesterol, for example, they can control the fat intake of their diet
- make sure they have all the appropriate tests such as mammograms and prostate gland tests. If the individual concerned is planning a family, their partner may also need to be tested to estimate the chances of children inheriting the condition. Unborn babies can be tested too

However, the results of these tests can cause problems: the knowledge that a foetus has a serious medical condition can lead to a difficult decision about whether to continue with the pregnancy, whereas the knowledge that an individual has the allele for Huntington's disease brings the likelihood of an early death. **Genetic counsellors** are there to give advice and support.

4 What is a DNA probe and what is it used for?

Answer on p. 112

Restriction mapping

Revised

When dealing with pieces of DNA such as cloned genes or plasmids, it is useful to know where the **restriction sites** are. A **restriction map** is a diagram that shows the position of the restriction sites of the relevant restriction enzymes. For example, to make a restriction map for the piece of DNA shown in Figure 10.4, we have to tag one end with a radioactive nucleotide and then digest it with a restriction enzyme. The resulting fragments can then be separated by **electrophoresis**.

Now test yourself

5 Define the following terms:
(a) recombinant DNA
(b) transgenic organism
(c) plasmid

Answers on p. 112

Tested

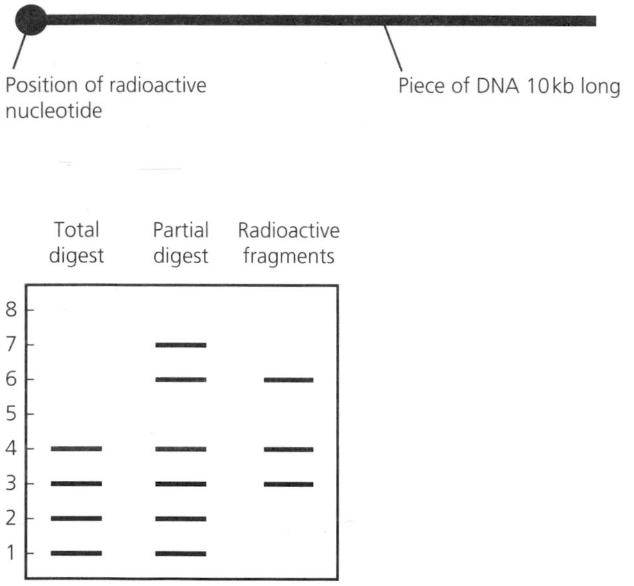

Position of radioactive nucleotide
Piece of DNA 10 kb long

Total digest
Partial digest
Radioactive fragments
8
7
6
5
4
3
2
1

Figure 10.4 A piece of DNA

To work out this restriction map, it is a bit of a logic puzzle:

- *In lane 1 — total digest*, we know that there are four fragments and so the enzyme must have made three cuts. We also know the fragments are 1, 2, 3 and 4 kb long, which means the original must be 10 kb long. However, we do not know the order of the fragments: it could be 1, 2, 3, 4, 1, 3, 2, 4 or any other combination.
- *In lane 2 — partial digest*, we know that the enzyme has not finished digesting (or 'cutting') the DNA because the pieces add up to more than 10, which is impossible. There must be fragments of 5, 6, 7, 8 or 9 kb which have not had all their restriction sites cut.
- *In lane 3 — radioactive fragments*, we know that the fragments are not completely digested because the lane contains some of the same fragments as lane 2. We also know that they must all contain the left-hand ends of the DNA fragment. The radioactive 3 kb fragment is the smallest one, so that must be first. There is also a radioactive 4 kb fragment, which can only be the 3 attached to the 1. We can also see a 6 kb fragment but no 8 kb, so the order must be 3 plus 1 plus 2.

Therefore, the restriction map for this piece of DNA is shown in Figure 10.5, with the * showing the restriction sites for that enzyme.

____3____*_1_*_2__*______4______

Figure 10.5 The restriction map

DNA sequencing

Revised

One way to find out the base sequence of a piece of DNA is by **Sanger sequencing**, also known as **chain termination** or **dideoxy sequencing**. By using slightly abnormal nucleotides, the process of DNA copying is halted because the DNA polymerase enzyme cannot continue.

To sequence a piece of DNA, you need four separate tubes. Into each tube add:

- many clones of the piece of DNA to be sequenced
- DNA polymerase
- primers
- all four normal nucleotides: dATP, dTTP, dCTP and dGTP
- just one abnormal dideoxy nucleotide: ddATP, ddTTP, ddCTP or ddGTP

In each tube, the DNA polymerase copies the target sequence until, by chance, it encounters an abnormal dideoxy nucleotide. At this point, the copying stops. By the law of averages, every position that can have an abnormal nucleotide will, sooner or later, get one. So, if your original sequence reads TGCAGGCAT, the DNA polymerase will add a strand that reads ACGTCCGTA unless it is cut short by an abnormal nucleotide (Table 10.3).

Table 10.3 DNA sequencing

Tube	Dideoxy nucleotide	Fragments of DNA in that tube (number on gel)
1	ddATP	TGCA (6) TGCAGGCA (2)
2	ddTTP	T (9) TGCAGGCAT (1)
3	ddCTP	TGC (7) TGCAGGC (3)
4	ddGTP	TG (8) TGCAG (5) TGCAGG (4)

If the contents of the four tubes are separated by electrophoresis, you can read off the sequence of bases (Figure 10.6). The smallest fragment travels farthest, so we know that T is first. The second smallest fragment is TG, so we know that G is second etc. In this way we get a sequence that is **complementary**, so it is easy to work out the original sequence.

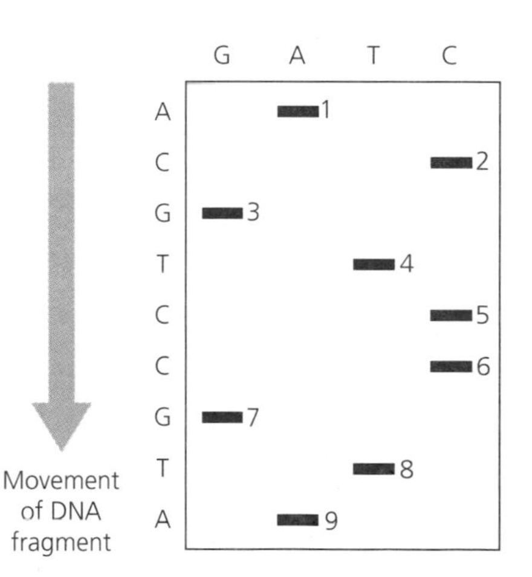

Figure 10.6 The sequence of bases can be read from the gel — we know that the original sequence must be complementary to the one we have here

Genetic fingerprinting

DNA profiling

Revised

The genes of different individuals are remarkably similar — they have to be in order to make vital proteins — but the non-coding DNA between the genes varies greatly and is unique to each individual. Specifically, there are many **repetitive, non-coding base sequences**. The probability of two individuals having the same repetitive sequence is very low. **Genetic fingerprinting**, or **DNA profiling**, shows up the similarities and differences between these regions (Figure 10.7). Thanks to the PCR, a DNA profile can be done on the smallest of samples, such as a speck of blood or one hair follicle.

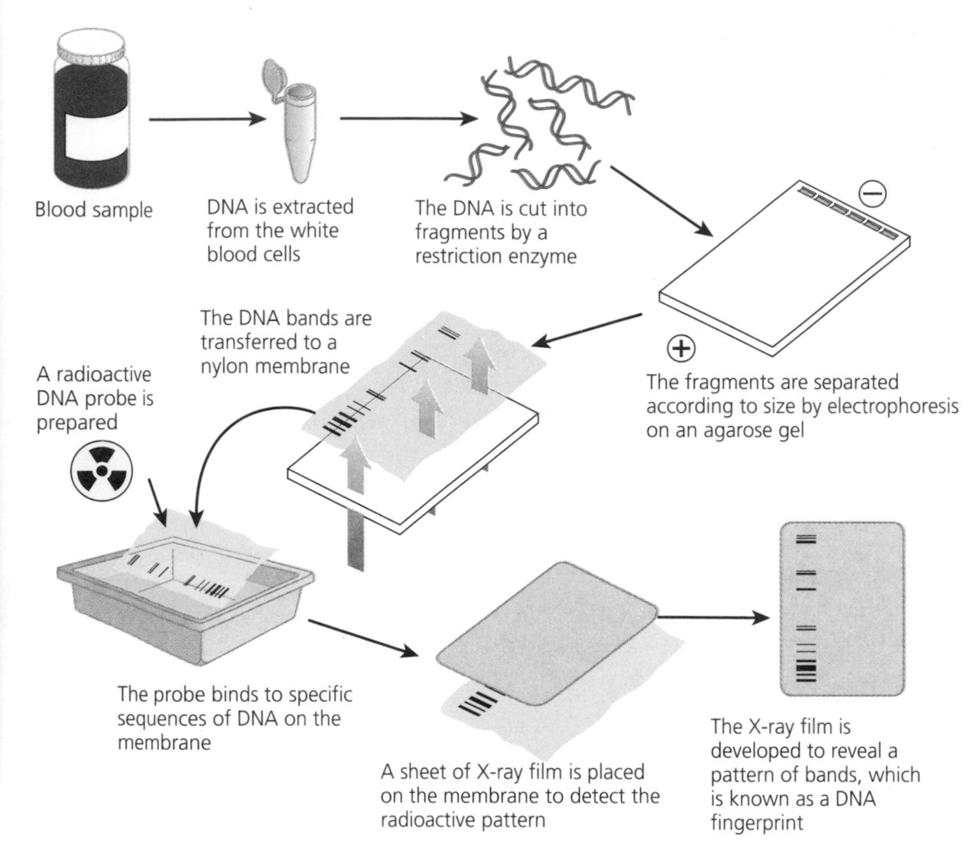

Figure 10.7 The basic stages of DNA profiling

Genetic fingerprinting can be used to:

- determine **genetic relationships** such as in paternity testing
- determine the **genetic variability** in a population
- ensure outbreeding — endangered species need as large a gene pool as possible, so DNA profiling ensures closely related individuals do not breed together
- establish pedigree in the case of thoroughbred racehorses
- identify crime suspects from evidence

Exam practice

1 The following diagram shows one type of plasmid used to insert a useful human gene (the gene of interest) into a bacterium.

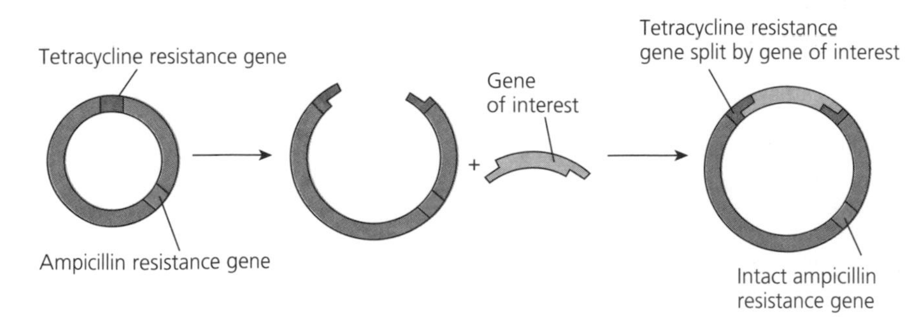

(a) Explain why the gene and the plasmid have to be cut with the same restriction enzyme. [1]

(b) When the enzyme ligase is added to the cut plasmids and the genes, two things can happen:

- the two sticky ends of the plasmids can simply rejoin, or
- the gene of interest can be inserted into the plasmid

Most bacteria do not take up the plasmid at all. The aim of the process is to find out which bacteria have taken up the plasmids with the human gene. Explain how scientists can use antibiotics to isolate the bacteria that contain the gene of interest. [2]

2 The following diagram shows the DNA profiles of a family with four children. The profiles were created by digesting an individual's DNA and separating the fragments by electrophoresis.

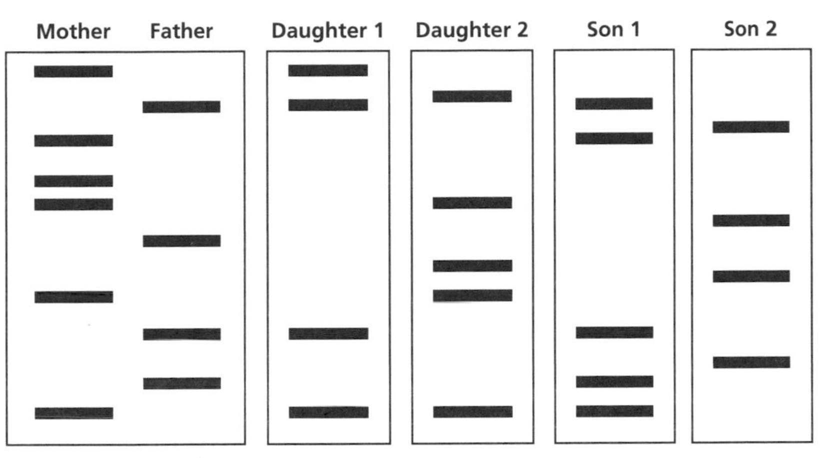

(a) Explain why DNA fragments move towards the positive terminal. [2]

(b) Explain why some fragments move faster than others. [2]

(c) Explain how the fragments are made visible. [1]

(d) We inherit our DNA from our parents and any fragments we do not inherit from one parent must come from the other. Use the profiles to determine the biological relationship of each child to their parents. [4]

Answers and quick quizzes online

Online

Examiner's summary

By the end of this chapter you should be able to understand:

- ✔ The three basic ways to obtain a fragment of DNA (make cDNA from mRNA, cut it out using restriction enzymes, clone it the polymerase chain reaction.
- ✔ The process of polymerase chain reaction (PCR).
- ✔ The process of using recombinant DNA technology to produce transformed organisms that benefit humans.
- ✔ The use of gene therapy to supplement defective genes.
- ✔ The use of labelled DNA probes and DNA hybridisation to locate specific genes.
- ✔ The process and uses of DNA profiling.
- ✔ The process of restriction mapping.
- ✔ The dideoxy method DNA sequencing.
- ✔ The use of DNA probes in genetic screening. The use of this information in genetic counselling, in family planning or medical treatment.

Now test yourself answers

Chapter 1

1 Population means all of the individuals of one species. Community means all the individuals of all species.

2 Biotic factors: predation (e.g. leaf-eating insects), disease, competition from other trees. Abiotic factors: any three from temperature, soil moisture, soil mineral ions, light intensity, pH, humidity.

3 **(a)** Quadrats (small ones, say 25 cm square).

(b) Belt transects.

(c) Interrupted belt transects.

4 $\frac{(200 \times 200)}{23} = 1739$ (rounded down)

5 Any three from better education, better diet, access to clean water, vaccinations, antibiotics.

6 A 2 million increase in 4 years is 500 000 a year. 500 000 as a percentage of 10 million is 5%.

7 **(a)** 55

(b) 73

(c) 83

8 **(a)** Poor diet; no vaccinations; waterborne diseases.

(b) Routine vaccinations; antibiotics; clean water.

(c) Better education and diet; early diagnosis of disease.

9 Temperature will have no effect on the growth of duckweed.

10 Any three from species/type of duckweed, number of plants at the start, mineral ions in the water, amount of light.

11 There is no correlation between the weight of the seed and the distance it falls from the tree.

Chapter 2

1 Chlorophyll, water, ADP and NADP.

2 ATP and NADPH (reduced NADP).

3 In photolysis, water is split to produce replacement electrons and oxygen is a by-product.

4 Glycolysis, the Krebs cycle and the electron transport chain.

5 **(a)** ATP is made in glycolysis and the Krebs cycle.

(b) ATP is made in the electron transport chain, powered by hydrogen ions.

6 The folded inner membrane (cristae) provides a large surface area for the reactions of the electron transport chain.

7 **(a)** 2

(b) 38

8 It is not reduced. It is reduced when it gains an electron, becoming NADH.

9 Volume of oxygen per unit mass per unit time, such as $cm^3\ g^{-1}\ m^{-1}$.

Chapter 3

1 Ringed seal.

2 Some is reflected off the surface, some misses the chloroplasts and some is the wrong wavelength.

3 Plant material contains more indigestible compounds than meat.

4 Herbivores need to eat constantly because a lot of their food is indigestible and contains relatively little energy. Meat contains a lot more energy. Carnivores can take in a lot of energy in a short time.

5 Snakes take in a huge amount of energy in one meal and do not have to expend energy in keeping warm.

6 Each transfer is inefficient, so the energy runs out.

7 Photosynthesis.

8 Respiration and combustion.

9 It has a strong triple bond that is difficult to break.

10 Turning nitrogen gas into ammonium; returning/adding it to the cycle.

11 By nitrogen-fixing bacteria or electrical storms.

12 **(a)** The plant gets nitrate, so it can grow in nutrient-poor soil.

(b) The bacteria are less likely to get eaten and they receive a supply of carbohydrate/sugar made by the plant.

13 The crop will grow without needing as much fertiliser, making it cheaper and leading to less environmental damage.

14 320 to 380 = 18.75% increase.

15 Increased carbon dioxide from respiration and combustion, deforestation which reduces photosynthesis; increased methane from anaerobic decomposition/cows traps more heat. All prevent heat being radiated out into space.

16 The aphid populations would be larger than normal, leading to more damage to crop/host plants.

17 An increased supply of ions leads to increased growth of algae.

Chapter 4

1 When a barren habitat becomes inhabited by pioneer species.

2 Existing species make conditions more favourable.

3 The final stage of succession: a stable community.

4 A mixture of rotting organic matter and bacteria that releases mineral ions slowly.

Chapter 5

1 An alternative form of a particular gene.

2 Zero.

3 Alleles which, if present, will both be expressed.

4 Red, pink, white in the ratio 1:2:1.

5 A or B.

6 Individuals 9, 11 and 14 are all born to non-albino parents. These parents must be carrying the albino allele without expressing it, which is the definition of a recessive allele.

7 A gene carried on a sex chromosome.

8 Females have two copies of each X-linked gene, males have only one copy. Therefore, in males, each X-linked gene is expressed because there is no second (dominant) copy to mask its effects.

9 Yes, but the mother must be a carrier and the father must be a haemophiliac, which is very rare.

10

Individual	Genotype	Reason
1	X^bY	He is a colour-blind male
2	X^BX^b or X^BX^B	She is a normal female who gave X^B to individuals 4, 5, 6 and 8. She may or may not have the second B allele
3	X^bY	He a colour-blind male
4	X^BX^b	She has a colour-blind daughter (individual 11), so she must be carrying the b allele
5	X^BY	He is a normal male
6	X^BX^b	She has normal vision, but has a colour-blind son (individual 12). Individual 7 has normal vision, so the b allele must come from her
7	X^BY	He is a normal male
8	X^BY	He is a normal male
9	X^BX^b	She has normal vision, but has a colour-blind son (individual 14). Individual 8 has normal vision, so the b allele must come from her
10	X^BY	He is a normal male
11	X^bX^b	She is a colour-blind female (quite rare)
12	X^bY	He is a colour-blind male
13	X^BX^B or X^BX^b	She is a normal female, but there is no way of telling what the second allele is
14	X^bY	He is a colour-blind male
15	X^BY	He is a normal male

11 All the different alleles present in a population.

12 They add up to more than 1.

Chapter 6

1 Any five from sound, light, heavy and light pressure, various chemicals (taste and smell), heat, cold, gravity, movement.

2 A plant growth response towards or away from an external stimulus.

3 Taxes are a directional response to a stimulus in organisms that can move. Kineses are a non-directional response to a stimulus in organisms that can move.

4 **(a)** They move towards the light.

(b) There is more light for photosynthesis.

5 To avoid predators, to avoid drying out and to increase the chance of finding rotting vegetation.

6 In a taxis, the organism would go straight to or from the light. In a kinesis, the organism would set off randomly in all directions, but slow down or stop when at its preferred side.

7 To prevent the brain being bombarded with too much information or to allow the body to focus on a new stimuli.

8 Visual acuity is the ability to see detail.

Chapter 7

1 Only the cells in the ovaries have the right receptor proteins in their membranes.

2 Local chemical mediators do not travel in the blood — they affect just the surrounding area. Hormones do travel in the blood and can affect target cells all over the body.

3 The cell body.

4 Axons take impulses away from cell body, whereas dendrons/dendrites take impulses towards the cell body.

5 Active transport and unequal facilitated diffusion.

6 −70 mV

7 The action potential arrives at the synapse → **calcium** ions flow into the pre-synaptic membrane → molecules of neurotransmitter **diffuse** across the synaptic cleft and fit into specific **receptor proteins** → The **permeability** of the post-synaptic membrane changes → **sodium** ions flow in, causing a positive charge to build up inside the **post-**synaptic membrane → If **threshold** is reached, an action potential is created in the post-synaptic neurone.

8 ATP is needed for the movement of vesicles, resynthesis of the neurotransmitter and active transport of calcium ions out of the synaptic knob.

9 If synapses were not unidirectional, the result would be chaos and there would be no coordination. Impulses could pass to sensory organs down sensory nerves and back from muscles down motor nerves.

10 The effects in order are inhibit, inhibit, inhibit, prolong, prolong.

11

Component	Role in contraction
Actin	One of the main proteins — the thin filaments
Myosin	The other major structural protein — the thick filaments that have many moveable heads
Troponin	Small, globular protein that binds to tropomyosin, moving it aside and exposing the myosin binding site on the actin
Tropomyosin	Long, thin fibrous protein that blocks the myosin binding site on the actin
ATP	Binds to the myosin head, splits to provide the energy to detach the myosin head and reattach further along
Calcium	Initiates contraction by activating troponin

Chapter 8

1 Negative feedback is the mechanism for stability/return to a set point. Positive feedback is the mechanism for change.

2 Endotherms maintain constant core temperature despite the environmental temperature. The core temperature of ectotherms is usually the same as the environmental temperature.

3

Hormone	Made by	Target organ	Action
FSH	**Pituitary gland**	Ovary	Causes a new follicle to develop
Oestrogen	Follicle	Various	Stimulates repair of the endometrium; combines with LH in a positive feedback to bring about ovulation
LH	Pituitary gland	**Follicle**	**Brings about ovulation**
Progesterone	**Corpus luteum**	Various	Maintains the uterus lining; inhibits FSH production

4 Progesterone inhibits FSH, so no new follicles develop and ovulation does not occur.

Chapter 9

1 They contain the base T, not U.

2

DNA sequence	AAT	CAT	GTC
mRNA sequence	UUA	GUA	CAG
Amino acid	Asparagine	Histidine	Valine

3 mRNA is variable in length, not folded and contains no hydrogen bonds. tRNA is a fixed length, is folded and contains hydrogen bonds.

4 There are 64 types because there are 64 different anticodons.

5 Pre-mRNA contains introns, whereas mature mRNA has had them removed.

6 186 (62 × 3).

7 The order is **(b)**, **(f)**, **(e)**, **(d)**, **(c)**, **(g)**, **(a)**.

Chapter 10

1 64

2 Sticky ends are staggered cuts in the DNA. They can be joined to complementary cuts made by the same enzyme.

3 Epithelial cells have a high turnover — they are lost and replaced constantly.

4 A short piece of labelled DNA that is complementary to a target sequence. It is used to find a particular gene/allele/base sequence.

5 **(a)** DNA from two different species joined together.

(b) An organism containing recombinant DNA.

(c) A circular piece of DNA found in bacteria.